63632
920 F36	FENTON, CARROLL L. STORY OF THE GREAT GEOLOGISTS 3.50

	63632
920 F36	FENTON, CARROLL L. STORY OF THE GREAT GEOLOGISTS 3.50

MAR 1 1 1997

San Mateo Public Library
San Mateo, CA 94402
"Questions Answered"

BRO DART PRINTED IN U.S.A.

These are stories of certain gifted men who sought to read the riddle of our earth in rocks and cliffs. From this account of their lives and discoveries we learn the truth about our planet; bit by bit the development of earth science unfolds in the light of its great moments and its great men.

In presenting biographical sketches of thirty-five leading geologists from the Greek, Aristotle, to the brilliant contemporary American, Chamberlin, the authors have emphasized the human side of this science. They tell, for instance, of Nicolas Desmarest, born in poverty, forced to make his expeditions on foot, living on dry bread and cheese, and sleeping on the earth floor of herdsmen's huts. Of the influential Werner, who spun theories by the fireside and isolated himself by refusing to read journals or open personal letters lest they disagree with his ideas. Of Louis Agassiz, who wrote "fiery discourses about a sheet of ice" and searched the world over for glaciers, leaving his wife to pine away and die. Such men were adventurers in the intellectual as well as the physical world and their stories are exciting and important.

Carroll and Mildred Fenton, both outstanding scientists, have the ability to translate scientific information into absorbing, popular reading. At present Mrs. Fenton is curator of the Geological Museum at Rutgers University and Dr. Fenton is science editor there.

BOOKS BY
CARROLL LANE FENTON AND
MILDRED ADAMS FENTON

The Story of the Great Geologists
The Rock Book
Mountains

BOOKS BY
CARROLL LANE FENTON

Our Amazing Earth
Our Living World

THE STORY OF
The Great Geologists

CARROLL LANE FENTON and
MILDRED ADAMS FENTON

Illustrated

DOUBLEDAY, DORAN AND CO., INC.
Garden City, New York
MCMXLV

920
F36

63632

COPYRIGHT, 1945
BY CARROLL LANE FENTON AND MILDRED ADAMS FENTON
ALL RIGHTS RESERVED
PRINTED IN THE UNITED STATES
AT
THE COUNTRY LIFE PRESS, GARDEN CITY, N. Y.
FIRST EDITION

Acknowledgment

Such a book as this always leans heavily upon others; we have tried to indicate our debt in the list of sources and references. Special acknowledgment, however, must be made to Mr. Noah T. Clarke for permission to reproduce the rare Swinton lithograph from his father's biography of James Hall. Also to the University of Pennsylvania Press, which permitted quotation from *Amos Eaton,* by Ethel M. McAllister, a work that furnished most of our information about the great educator of Troy.

Several portraits have been taken or modified from those in standard biographies and are so acknowledged, as are photographs from other sources. The portrait of Hutton is based in part upon photographs of a draped bust and partly upon paintings which show the subject in late eighteenth-century clothing. One of these is familiar as the frontispiece of *The Birth and Development of the Geological Sciences,* by F. D. Adams. Plates in this work also furnished a basis for the portraits of Werner and Von Buch.

Carroll Lane Fenton
Mildred Adams Fenton

FOREWORD

From Sandbank to Science

WE WALKED, late one afternoon, up a rock-rimmed trail that climbed from Arizona's Grand Canyon. Behind us lay the Inner Gorge, its walls descending in darkly contorted masses to the turbulent Colorado. On our right and left spread olive-ochre slopes, rising to precipitous walls of red, buff and rusty white stone. These walls stretched away for miles—a maze of cliffs, re-entrants and promontories along which the yellowing sunlight glowed, or glanced from turrets and ridges based in purpling haze. Those near us seemed to wax and shift as moving rays picked them out or allowed deepening shadows to grow from intervening clefts.

Such changes of light and scene held drama, yet they lacked the solid stuff of meaning found in the canyon itself. The forbidding Inner Gorge, for example: its dark schists with their tortuous laminae tell of an early revolution which compressed a segment of the earth and crumpled it into mountain ranges as molten granite oozed upward into every fracture and gap. A line shows that these mountains were worn away—reduced to a bare, almost level lowland that was flooded by an arm of the sea. Buff strata, however, recall ancient sandstorms, while brick-red ledges tell of pools where awkward amphibians came to mate in the stir of long-forgotten springs. The canyon rim

Foreword

again tells of uplift that raised low basins into plateaus and gave new vigor to what had been leisurely, rather futile streams. So strong did one, the Colorado, become that it has carved this most magnificent of canyons in less than two million years.

Records of these events are not secrets; they lie in bold cliffs and ledges for all who look to see. Yet their story went unread until 1869, when the first explorers guided battered boats seaward through the Inner Gorge. The men were hungry, exhausted, worn by dissension; their leader bore added burdens of pain as well as disappointment that grew as instrument after instrument went overboard or was soaked and pounded into uselessness by the raging stream. Yet doggedly he continued work, following ledges, probing tributaries, climbing cliffs, or making such observations as remaining equipment allowed. In the end he emerged with an initial and surprisingly adequate account of the story told in chasms and cliffs.

This explorer discovered the Grand Canyon's records; he alone did not learn their meaning. That task had been carried forward through centuries, since a learned Greek profiteer first noticed that waves erode cliffs and that rivers, by building deltas, add new land to old. Other men watched volcanoes erupt and attempted to explain them, traced uplift and subsidence in seashores, or discovered that streams could both erode valleys and shape mountains of firm, resistant stone. At last came a band of clergymen, teachers, officials, philosophers and men of leisure who turned what had been mere scattered, poorly organized observations into a science of the earth. In it were data, principles and hypotheses to serve and guide later explorers, as well as trained scientists who in time would carry geology back through earth's early ages or project it across space and eons to planets sprung from a tidally disrupted sun.

From sandbanks to solar system; from waves that beat on modern shores to crystalline, contorted rocks that settled as mud in seas of two billion years ago. The story is long, diverse, complex, yet not too much so to be told. We have chosen to

Foreword

tell it through the lives of men—men who in their diverse ways loved our planet and labored to make it known. Their successes became the stuff of earth science; their failures show how it forged forward though impeded by ignorance, adverse times or human fallibility.

For men of science are human; let no one be in doubt upon that. No man is transformed into a flawless mental machine by mounting to a professorial chair, by writing a learned treatise, or by joining a governmental survey. With him in his rise go hereditary traits, effects of early training or mishaps, seeds of organic ills. Each of these will bear upon his work, as will thought patterns within which he has matured and forces that stimulate or restrict the society to which he belongs. Such factors mold both men and their thinking, and must be taken into account when we tell how a science has grown.

Even then the account cannot be complete—not, at least, for earth science in a book no larger than this. Geology once was compact, simple, unified; a body of knowledge about our earth as seen by man's unaided eyes. But that pristine stage was passed so long ago that an American geologic pioneer had to call a man of microscopes from Europe to aid in understanding our West. Today the science has become so much divided that men who work in one field may neither know nor understand those who specialize in another. Indeed, the task of correlating results from widely varied branches has become a specialty in itself.

The growth of these departments might, of course, make a book: a large and doubtless technical treatise written by authorities for their learned scientific peers. But ours is not a reference book, nor is diversity our chief concern. In a world where all things from politics and paint to hardened magmas are related, we prefer to write about men who dealt with the earth as a whole, or with features of such magnitude that they influence all geologic thought. From such investigators came the modern picture of our planet, with its sweep of geologic change that rises from an ancient past and sets a frame for the future.

Foreword

It is a picture made strong and clear by understanding, yet fired by the light of inspiration that comes with discovery. That light shines dimly through ancient and oft-copied books, glows more brightly in the work of men who built a science among hedgerows and alpine valleys of Europe, and reaches its full intensity with those who tried and matured geologic knowledge in a growing and stimulating New World. For there the men of rocks became pioneers, journeying into unknown regions where they shared the glory of voyageurs and frontiersmen, of La Salle, Frémont and De Smet. Wresting facts and viewpoints from a vast wilderness, they achieved stature commensurate with the country and set a precedent for bold explorations which now penetrate depths of land and sea, reach into polar wastes, and cross circumstellar space to find an origin for our planet.

The earth itself has adventured greatly, and so have the men who probed its past or explained its present changes. Their physical exploits were matched by mental; by daring thought that replaced legend with observation, that overcame prejudice, that supplanted medieval fear of the world with modern understanding. Greatest of all, perhaps, was the development of that regard for truth which enables men to give up familiar but false knowledge and even to enjoy finding error in their own discoveries.

These triumphs have given human beings a new, an improved world outlook; they have added to our joy in the world; they have helped us build up stores of knowledge with which we make use of the earth and improve our lot upon it. We owe our ways of living, if not life itself, to the men who achieved this revolution. Need we have a better reason to know them, to trace their exploits as they conquered frontiers of ignorance, wilderness and human weakness to make earth's story known?

Contents

		PAGE
FOREWORD: FROM SANDBANK TO SCIENCE		vii

CHAPTER
I	FLUIDS AND EXHALATIONS	1
II	GROPINGS SUCCEED DECAY	16
III	MAPS AND ANCIENT VOLCANOES	27
IV	GEOLOGY BY DICTUM	39
V	THOSE SKEPTICAL SCOTS	49
VI	NEPTUNE VERSUS VULCAN	61
VII	LIKE GOES WITH LIKE	70
VIII	KNIGHT OF THE "PRINCIPLES"	84
IX	THE CAMBRIAN CONFLICT	98
X	AGASSIZ OF THE ICE AGE	111
XI	NEW SCIENCE TO NEW WORLD	124
XII	EX-CONVICT PROFESSOR	137
XIII	GEOLOGIST AT LARGE	150
XIV	IMMIGRANT INNOVATOR	165
XV	THAT A NATION MIGHT GROW	179

Contents

CHAPTER		PAGE
XVI	The Canadian Shield	191
XVII	Earth's Changing Time Scale	208
XVIII	Railroads and New Surveys	216
XIX	Canyon's Conqueror	232
XX	Earth Blisters and Changing Land	251
XXI	Glaciers to Galaxies	270
Sources and References		287
Index		295

Half-tone Illustrations

Facing Page

ABRAHAM GOTTLOB WERNER	14
LEOPOLD VON BUCH	14
JAMES HUTTON	15
CHARLES LYELL	30
WILLIAM SMITH	30
PART OF WILLIAM SMITH'S GREAT MAP	31
ADAM SEDGWICK	94
SIR RODERICK MURCHISON	94
LOUIS AGASSIZ	95
WILLIAM MACLURE	95
BENJAMIN SILLIMAN	110
SAMUEL L. MITCHILL	110
AMOS EATON	111
JAMES HALL	174
EBENEZER EMMONS	174
DAVID DALE OWEN AND A WOOD ENGRAVING SHOWING A CAMP OF HIS ASSISTANT EVANS	175
SIR J. WILLIAM DAWSON	190
TWO SPECIMENS OF *EOZOON*	190
SIR WILLIAM E. LOGAN	191
SOME OF THE CRUMPLED MARINE ROCKS WHICH HE STUDIED ON THE CANADIAN SHIELD	191

Half-tone Illustrations

Facing Page

CLARENCE KING	254
JOHN WESLEY POWELL	254
FERDINAND HAYDEN	255
ONE OF HIS FIELD PARTIES IN CAMP AT RED BUTTES	255
GROVE KARL GILBERT	270
THOMAS CHROWDER CHAMBERLIN	271

Linecut Illustrations

	Page
A Tongue Stone and Two Cerauniae	21
Basalt Columns	35
Title Page of Hutton's Great Work	55
Diagram Showing How William Smith Used Fossils to Match Beds in a Quarry, a Hill and a Canal Cut	74
The Rev. Benjamin Richardson	76
The Rev. Joseph Townsend	77
Sedgwick on a Field Trip	102
Sedgwick in Academic Costume	108
Map Illustrating Volney's Theory of a Great Ancient Lake	127
Salary Check from James Hall to Charles Schuchert	159
Proof That Hall Wrote His Own Manuscript Long After He Engaged Assistants	160
Limestone with Specimens of *Archimedes*	171
Owen's Explorations in the Northwest	176
Logan's Sketch Showing One of His Camps	183
Scene on the River Rouge	188
Pre-Cambrian and Canadian Shield	195
Chart of Pre-Cambrian Times and Rocks	201
North America in Late Proterozoic Times	203
Geologic Time Chart	212–15

Linecut Illustrations

	Page
Principal Routes of the Pacific Railroad Surveys of 1853–56	218
Explorations of Early Surveys	227
Powell's Route Down the Green and Colorado	238
Jukes Butte in the Henry Mountains	257
"Ways and Means"	259
Lake Bonneville	261
Bonneville and Other Ice Age Lakes	265
Lobes of Two Continental Ice Sheets	275

THE STORY OF
The Great Geologists

CHAPTER I

Fluids and Exhalations

SUNSHINE STILL FIRES the Grand Canyon's walls, but they and the turbid brown river now are far away. We sit in the cool north light of a library, scanning maps that recall those early days when the whole occidental world was no more than a fringe surrounding the Mediterranean Sea. Into that sea stretched Greece, or Hellas, across trade routes that ran east, west, northeast and south. Some passed between peninsulas and islands, where salt water has so far divided and conquered that only a few parts of the country lie more than fifty miles from the sea.

Land between straits and embayments is rough: a complex of mountain chains in which rocks were raised, crumpled and broken by movements of earth's crust. Their heights rise to rugged Mount Olympus, on whose 9700-foot peak the gods were supposed to dwell. Between ranges lie valleys with intermittent streams which turn the dry channels of summer into foaming torrents when winter rains begin. A few rivers plunge into caverns, to reappear as giant springs that gush from crumbling cliffs.

The country is one of great contrasts within short distances,

The Story of the Great Geologists

though almost any distance must have seemed great in the period recorded by our maps. Hemmed in by rocky heights and salt water, the dwellers in each group of valleys clung together, developing their own dialect and culture, and their own economic life. All centered in one or two small cities with their tributary hamlets and farms, and a hinterland of grazing country that stretched up the least precipitous slopes. There uncouth herdsmen guarded flocks of thin, short-fleeced sheep, unaware that sentiment would one day transform them into the graceful, pipe-playing shepherds of ballet and poetry.

Such was the Greek city-state, a product of human settlement amid geologic barriers. Since mountains stubbornly resisted travel, the more ambitious townsmen finally learned to make long journeys by sea. Building ships upon designs borrowed from older peoples, they set out on voyages of commerce as well as conquest. Before 600 B.C., for instance, merchant sailors of Miletus (a city-state in Asia Minor) had captured coastwise trade to the Black Sea and had established a shipping town at one of the Nile's many mouths. There they took on and discharged cargoes, bought and sold, and even ran a scarab factory that produced sacred beetles of clay for middle-class Egyptians.

In Miletus itself, as one of its merchants, lived a learned aristocrat named Thales. Born about 636 B.C., he apparently taught himself such learning as Greece then afforded and made a business trip to Egypt, where priests schooled him in astronomy and mathematics. Returning to Miletus, he achieved wealth—rumor says by cornering the market on mowing implements just before a bumper harvest. He also gained fame by predicting an eclipse of the sun that was to fall on May 28 in the year 585 B.C. Coming in the midst of a battle between the Medes and Lydians, this eclipse so awed both sides that they declared a truce which soon led to peace.

Like many another ancient wise man, Thales was more at home with angles, ellipses and heavenly bodies than with processes at work on the earth. But he did watch waves as they

Fluids and Exhalations

crashed against shores with violence that shook the ground and broke rocks to pieces. He also examined mudbanks at mouths of the Nile and saw the Meander River pile up silt in the harbor of his home city. From such sights he drew an important conclusion: that water could change our planet's surface either by wearing shores away or by building them seaward. Though doubted and ignored again and again, that conclusion still stands as a fundamental in geology.

Thales was less successful when he tried to explain earthquakes. Waves, it was plain, could shake the land; during great storms the islands near Miletus appeared to rock like ships. The implication seemed obvious: the whole earth must be compared to a ship riding on an encircling sea. When its waters tossed and streamed with sufficient violence they were bound to shake the rocky "vessel" and so produce quakes of varying intensity.

Thales seemingly became obsessed by water, finding in it the source of all things both lifeless and animate. His student Anaximander disagreed, imagining an all-pervasive primeval substance from which movements as well as things were derived. But common men did not rise to such speculative heights, being content with established goddesses and gods who upset nature at will. Thus the mountain-rimmed plain of Thessaly was explained as a one-time lake drained by Poseidon, who shattered ranges to make the gorge of Tempe, through which lake waters rushed to the sea. Even the widely traveled Herodotus clung to that tale; writing about 440 B.C., he remarked, "That the gorge of Tempe was caused by Poseidon is probable; at least, anyone who credits earthquakes and chasms to that god would say this gorge was his work. To me it seemed quite plain that the mountains there had been torn apart by an earthquake."

Nor was Herodotus much less naïve when he tried to explain fossils. Anaximander had seen fish remains in ancient strata and—possibly parroting Egyptian scholars—had said that they once were alive. A century after Anaximander's

The Story of the Great Geologists

death, the exiled Xenophones of Colophon described fossil fish in Sicilian quarries near Syracuse and petrified shells in uptilted rocks of the Italian mountains. Both, said he, proved that seas once covered what now is dry land, and Xanthos of Sardis agreed when he found shells in the hills of Asia Minor. Yet when Herodotus saw remains of one-celled creatures in Egyptian limestones, he concluded they must be petrified lentils left from stores of food for slaves who had built the Pyramids.

While knowledge thus progressed and receded, both Greece and the Greek world changed. Miletus ceased to be important; leadership crossed the Aegean, where Athens and Sparta shared dominance and engaged in constant quarrels. In Asia the Persian Empire forged ahead, finally invading Hellas. Sparta and Athens stopped bickering long enough to defeat the invaders, fighting Darius, Xerxes and Artaxerxes in a war that began before Herodotus was born and dragged on till 449 B.C., when he was thirty-five.

It is said that in this struggle Sparta furnished the army and Athens the navy, a statement that is very nearly true. When the fighting ceased, Sparta disbanded her troops and began to suffer the twin evils of isolation and unemployment. Athens, on the other hand, turned its warships into a merchant fleet, becoming a trading city of first rank and a meeting place for able and inquiring minds of the whole Mediterranean basin.

Some Athenians chose to build fortunes; others wrote, sculptured, philosophized or explored the mysteries of nature. Under the statesman Pericles, they gave the city an intellectual vigor that survived almost thirty years of war with an envious and resurgent Sparta, followed by decades of intrigue and conflict between leagues of city-states. A declining and defeated Athens still was the cultural capital of Europe in 335 B.C., when a Macedonian scholar named Aristotle set up a school in the city's most fashionable gymnasium.

Aristotle was born at Stagira, some two hundred miles north of Athens, in the year 384 B.C. His father was a rich and

Fluids and Exhalations

aristocratic leader, physician to the Macedonian king. The son liked medicine and learned much of its lore, but tales that he dissected human bodies appear to have no basis. Throughout his life Aristotle showed profound ignorance of anatomy, even miscounting the number of man's ribs and calling the brain a bloodless organ which did not reach the rear of the skull.

His father died when Aristotle was only seventeen years old. There are rumors that the youth squandered his fortune on liquor and gambling, joined the army to avoid starvation, left it to sell drugs, and at thirty underwent a reformation which took him to Athens for study under Plato. All this seems to be mere envious scandal. A more probable account sends Aristotle to Athens at the age of eighteen, ready for a game, a girl or a drink, but with enough wealth to pay three Attic talents (more than thirty-five hundred dollars) for a set of books on philosophy. Finding that Plato was abroad, he also showed enough intellectual drive to study hard for three years while awaiting the master's return. Those years probably removed most of the inexperience, rugged manners and northern accent which must have seemed uncouth to smug Athenians. Indeed, it is said that Aristotle soon outdid them in polish, becoming foppish, overnice and lispingly affected in speech.

If we accept this account, Aristotle studied under Plato seventeen years and developed into an able and respected, though not a favorite, pupil. He was short, slender and delicate, with uncertain health and a skill with biting phrases that made many enemies. He also was interested in stars, animals, plants, weather, earth—subjects with no charm for Plato, who yearned for a perfect aristocracy and an all but perfect state. Teacher and pupil were bound to drift apart. During Aristotle's last years of study he gave independent lectures and sometimes criticized Plato's ideas, though the two men seemingly did not quarrel.

Plato died in 347 B.C. and was succeeded by his nephew. Aristotle went to Asia Minor, where he taught three years at Atarneus and married the ruler's adopted daughter. Their

The Story of the Great Geologists

honeymoon was spent on Lesbos, whose rocky shores with their inlets and tide pools gave Aristotle a chance to study marine organisms. He discovered that sponges were animals, watched crabs and snails, and studied the sense organs of scallops, which guide themselves by light and shade as they skip through shallow water. His studies were interrupted by messengers from King Philip II, asking him to become tutor of the young Prince Alexander. Aristotle, now a widower, accepted and sailed for Philip's capital.

The Prince was thirteen or fourteen, precocious and readily impressed by a man of tremendous learning. But he also was undisciplined, vain, epileptic, with emotions incurably warped by a barbaric and superstitious mother. He mixed schooling with politics and court intrigue, reigned in his father's place at the age of seventeen, and at eighteen helped to conquer Athens. Aristotle moved to Stagira, where he wrote and taught a few select pupils. Philip meanwhile was assassinated and Alexander took the throne.

Aristotle must have known the new King's faults, but was not much disturbed by them. He despised the fraudulent democracy of Athens; knowing what Hellas had lost by internecine strife, he was ready to support a monarch who carried forward Philip's plan by uniting Greece, western Asia and Egypt into one compact nation. Aristotle is supposed to have furnished Alexander with manuals of imperialism; in return he received royal funds which have been variously estimated at the equivalent of a half million, two million, or four million dollars. Three centuries later, Pliny said that Alexander ordered royal gamekeepers, fishermen and gardeners to provide his former teacher with animals and plants; other ancient writers say that Aristotle once had a thousand collectors scattered through Greece and Asia Minor. Royal gifts probably continued after he went to Athens, for no private fortune could have bought his library nor paid the assistants who gathered data for his three or four hundred books.

Alexander at last withdrew his favor when a nephew and

Fluids and Exhalations

pupil of Aristotle would not worship the monarch as a god. The stiff-necked nephew died in jail; his uncle was let off with a hint that even great thinkers might find it fatal to censure the doings of kings. Then Alexander died after a drunken orgy, Athens revolted, and patriots who hated all Macedonians declaimed against Aristotle. Soon they dragged out the charge of impiety—the same trumped-up batch of nonsense that had done Socrates to death. Rather than face a hostile mob, Aristotle chose exile at Chalcis and there in 322 B.C. he died.

Aristotle was no scientist, for scientists did not exist in the fourth century B.C. He was a philosopher whose field was the universe, a teacher whose mission was to eradicate ancient error while passing on the truths he had learned. We can picture him as he lectured: a short man in immaculate, flowing garments who strolled along a covered walk or paused to unroll some clumsy book brought by a plodding assistant. Students clustered about, listening intently or scribbling notes on plates of wood covered with blackened beeswax.

Sometimes the talk was of logic; of how to reason so perfectly that error could not creep in. At other times it ranged stellar spaces, dealt with the forms of animals, or spun fine arguments about dreams and the soul. Only now and then did it touch the earth, which in spite of some ancient errors seemed to offer fewer puzzles than did life, art or politics.

Those brief comments on our planet were not organized; we must get them from a polyglot treatise on what then was termed meteorology. In it (or in the course for which it probably formed lecture notes) Aristotle taught that the earth was a globe smaller than most stars, yet center of the universe. The interior of the globe was cavernous and hot; its surface was sometimes shaken by earthquakes and sometimes devastated by volcanic eruptions. The former had been attributed to water, uprushing ether, and collapse of the spongy interior, but such explanations were superficial and false. The real cause of earthquakes, said Aristotle, was wind produced by

The Story of the Great Geologists

evaporation of moisture either on or within the earth. Small shocks might come from wind pushing outward, but great ones were made by gales that rushed into subterranean caverns or were forced back into them by waves.

This was said with a good deal of aplomb, as modern lecturers sometimes announce the "latest opinion of science." Yet we soon trace it back to the legend of Aeolus, who imprisoned winds in a cave beneath volcanic islands where eruptions have taken place since the dawn of history. Aristotle apparently recognized some connection between volcanoes and earthquakes, for he described one outburst in which "the earth swelled up . . . and a mound arose with a noise; finally the mound burst, and a great wind came out of it, throwing up live cinders and ashes."

The heat of eruptions must come from fire, but Aristotle smiled at those who thought that the earth's interior raged with primeval conflagrations. Fires at the surface could be made by friction; those underground were produced anew as air that whistled through caverns was broken to bits which then were beaten about until they burst into flames. Hot springs, often found near volcanoes, came from sea water similarly heated as it flowed through underground channels which finally brought it back to the surface.

In Aristotle's time, as today, little was known about rocks deep down within the earth. But thousands of slaves were busy quarrying stones and digging out minerals that afforded metals or gems. The philosopher explained these substances by means of two exhalations which arose from our planet's interior as well as from its surface. The dry, or smoky, exhalation might be seen in deserts; its heat produced ordinary stones as well as sulphur, red iron ore, and other earthy minerals. The moist exhalation, or vapor, resembled steam; when imprisoned in cracks and small holes, it left deposits of iron, copper, gold and other pure metals. Attempting to classify these substances, Aristotle could only conclude that they were "water in a sense,

Fluids and Exhalations

and in a sense not. Their matter is that which might have become water, but it can no longer do so."

Though moist vapor came from within the earth, Aristotle ridiculed the notion that rivers must flow from one great central mass of liquid or from subterranean lakes. True, water "raised" by the sun came down again as rain, soaking into reservoirs by which some streams were fed. Most rain, however, fell upon mountains, where it mingled with water that had "existed as such" as well as with liquid produced by condensation when air within the ground became cold. These three sources provided enough moisture to saturate highlands until they resembled huge sponges from which water oozed or trickled to the surface, where it formed the headwaters of streams.

Was there proof for this theory? Plenty of it. When workmen dug aqueducts, liquid collected in the ditches "as if the earth in its higher ground were sweating the water out." Springs were found among mountains and hills; rivers flowed from highlands and the largest streams always came from the highest peaks. Aristotle described the sources of many rivers, including the Nile, the Don and the Danube, and found them all in mountain ranges. Incidentally, he made the Danube rise in the Pyrenees and the Don in the Hindu Kush Mountains of central Asia, though the latter's true source is a Russian lake only five hundred feet above sea level. He evidently could see no direct connection between rain or melting snow and floods, since water which fell upon the earth must percolate or drip into streams instead of running to them on the surface.

Like Thales, Aristotle had seen rivers deposit mud and knew that stream-laid silt had formed Egypt's fertile plain and its steadily expanding delta. The last of these changes, he thought, had far-reaching effects, since salt water pushed from the coast of Egypt must overflow that of some other land.

Deltas were built up slowly; so slowly that long periods must pass before they turned sea bottoms into land and caused the inundation of other regions. But Aristotle had no skimpy

The Story of the Great Geologists

ideas of time, and with eternity as a background he developed a theory of cycles during which silt built lands that then were flooded, while submerged areas "of necessity" were left dry again. The cause of those cycles did not lie in the earth, but in fluctuations of cold and heat "which increase and diminish on account of the sun and its course. It is owing to them that the parts of the earth come to have a different character—that some parts remain moist for a certain time, and then dry up and grow old while other parts in their turn are filled with life and moisture."

We can almost see pupils give puzzled frowns as they tried to reduce this speculative theory to clear and orderly notes. As if to help them Aristotle repeated, comparing the earth to a human body which grows, matures and at last sinks into old age. But earth's aging affected only special parts and in these it was bound to be terminated by processes of rejuvenation. Here, perhaps, some student from another school expressed doubt, for Aristotle burst out in condemnation of people who cherished "narrow views" and who doubted that the universe could remain indestructible yet go through extensive changes. Time fails not, the universe is eternal, but rivers and seas are temporary. Assuredly, then, lands and seas alternate and rivers—which have not flowed forever—are bound to come to an end!

The explanation is more involved and dubious than the statement, but Aristotle's theory simmers down to this: because of unspecified changes in the sun and its travels, parts of the earth go through climatic variations like those of winter and summer in Greece. During the former rainfall becomes abundant, so that highlands soak up more and more water, making large rivers that carry increasing quantities of silt. Water coming from streams makes seas deepen, while silting compels them to spread over land that is not receiving additions. In time, however, the cycle swings to its "summer" stage, in which the moist climate yields to aridity that parches the land, makes rivers shrink, and causes the sea to recede.

Fluids and Exhalations

But since earth's total rainfall does not change, a moist cycle must at once begin elsewhere in a region that once was dry.

When Aristotle fled from rabble rousers a favorite pupil, Theophrastus, took his place in Athens. Though also an outlander, Theophrastus was popular, managing the school with such success that it sometimes had two thousand students. He ranks, too, as the world's first important popularizer of science, for his works on plants—eighteen books in all—were widely read, justly admired, and remained the foremost botanical treatises for some eighteen hundred years. A work *Concerning Stones* was almost as popular for, dealing with both minerals and rocks, it contained practical facts secured from miners, quarrymen, metalworkers and jewelers. Though only fourteen pages of it have been preserved, it apparently survived for at least three centuries and was one of the sources used by Pliny when he wrote his *Natural History*.

The three centuries between Theophrastus and Pliny witnessed great changes in the Mediterranean world. Alexander's jerry-built empire crumbled while a vigorous new power, Rome, appeared west of the Adriatic. It conquered Carthage, defeated Macedonia, and then moved southward to "deliver all Hellenes from the Macedonian yoke." But the city-states used their newly gained freedom for squabbles which Rome settled by conquest. Corinth was burned, the Proconsul of Macedonia became Governor of Greece, and even philosophers were enslaved. Rome itself changed almost as much, for the conquering republic became an empire in which rich men grew richer, poor men poorer, and learning (guaranteed Greek and authentic) was imported from Athens or Asia.

Strabo was an Asiatic Greek who thought well of his Roman rulers and wrote a geography for their use. Like Aristotle, he believed that all things must change, and he used the Nile's flood plain and delta as examples of that truth. He also was a keen observer—too keen to accept Aristotle's notion that lands merely stayed where they were or were built upward while seas rose and sank. Instead he described quick upward movements accompanied by earthquakes, as well as slow ones that raised wide stretches of land without perceptible disturbance.

Uplift was caused by internal fires, said Strabo, working in ways not specified. Earthquakes, however, came from subterranean winds which also carried heat and flames to the surface through volcanic craters. Strabo described Mount Etna as well as Vulcano itself; from the former he watched streams of lava roll, looking like hot black mud that hardened into stone. He recognized Vesuvius as a volcano with rich, cindery soil spread over its slopes. In another field, he realized that rains caused rivers to swell and overflow, though he thought their source was water that existed as a primitive part of the earth. Those original supplies kept the streams going during periods of drought.

Strabo traveled widely; as far as the knowledge of his time permitted, he also read critically. Neither trait appeared in Pliny the Elder, hailed by Romans as a tower of wisdom whose books not only were numerous but "contained much abstruse matter."

Born about A.D. 23, soon after Strabo's death, Pliny was a native Roman of rich and aristocratic family. After a short career as a lawyer he went soldiering in Germany, served long terms as an official, and at last received command of a fleet based not far from Mount Vesuvius. There in A.D. 79 he saw the eruption that destroyed Pompeii and several near-by towns. Going to rescue friends and watch the disaster, the portly prefect made a great show of indifference to danger. While others tied pillows on their heads to ward off falling stones, he stopped, lay down on a clean sailcloth—and promptly was overcome by a blast of sulphurous gas. His body was recovered three days later "without any marks of violence upon it . . . and looking more like a man asleep than dead."

Though a man of affairs, Pliny was bookish; one of those earnest and overserious souls who delight in reading just for the sake of learning what has been put into books. He slept

Fluids and Exhalations

briefly and began work at two o'clock or sometimes at midnight, thus gaining a few hours before reporting to his monarch, a task he carried out before daybreak. Books were read to him while he rested and ate; he grumbled at the slightest loss of time and covered thousands of wax-covered tablets with notes. "No book is so bad," he liked to say, "that it does not contain something of value to me."

Chief fruit of all this owlish labor was the *Natural History*. Pliny himself boasted that it contained "twenty thousand things all worthy of regard" culled from two thousand volumes by at least six hundred and seventy authors. Actually, the work is an encyclopedia of fact, theory, hearsay and falsehood, all put together with little order and no trace of critical caution. With equal earnestness Pliny described the ways of oxen, horses, unicorns and dragons; he told how elephants were trained and quoted one Juba, King of the Mauritanes, "renowned for his studies and love of good letters," who maintained that whales 600 feet long and 360 feet wide shot themselves out of the sea into rivers of Arabia. After saying that fresh goat's blood robbed the diamond of its hardness, Pliny stated the general rule that underlay such credulity:

I would gladly know whose discovery this might be to soak the diamond in goat's blood; who first thought of this; or rather by what chances it was found out and known. . . . I must ascribe this discovery and all others like it to the might and beneficence of the divine powers. Neither are we to argue and reason how and why nature has done this and that; sufficient it is that her will was so and thus she would have it.

Coming to the earth, Pliny called it the suffering "mother of all," which causes earthquakes in protest against the wickedness of men who mine ores of gold, silver and iron. Having thus sentimentalized and exclaimed, he told where metals were found, what they were used for, and how some of them were mined. He thought that iron ore constantly grew upon the island of Elba but said that the greatest deposits were found in Vizcaya, Spain, where important mines are worked to this day. He denied that most tin came from "islands of the Atlantic Sea"—Britain—and described marbles, building stones and other rocks in daily use at Rome. He gave a detailed account of quartz but insisted that it was hard ice which grew in the North, upon high peaks, and on an island near Arabia. He thought emerald colored the air around it and that amber gave protection from poison, as well as from "illusions and alarms that drive people out of their wits."

Earth-shaking winds, exhalations, rocks that grew like plants—such contributions to earth science were cherished in ancient times. We might easily dismiss the lot with a shrug for Aristotle's assumptions and a smile for the tall tales retold by credulous Pliny.

Tolerant dismissal would be easy, but it also would be unfair. After all, the ancients *were* ancient: men close to myth and folk knowledge, in a world where almost everything was unknown. They had no store of authenticated data, no experience to help them tell fact from fancy or assumption. Fossils brought to Herodotus looked like lentils; more than two thousand years must pass before they were found to be protozoans, and even now there is doubt as to just what protozoans are. Aristotle saw waves trap air in coastal caves; for him they drove it into deeper caverns whose existence was axiomatic in 335 B.C. Neither Strabo nor any other ancient could have shown that rain water which sank into the ground might move through it for years or centuries before reaching the surface. It seemed obvious, therefore, that springs which flowed during long droughts were fed by supplies of liquid hidden within the earth.

We strive to realize these limitations for they, rather than modern knowledge, show how far ancient thinkers progressed. On one hand they gathered facts, recording sound observations on rivers, deltas and flood plains, earthquakes, volcanic eruptions, upraised strata, and certain fossils. On the other they began the long struggle to oust magic and capricious folk gods, substituting natural causes in affairs of the earth. We find

ABRAHAM GOTTLOB WERNER
first great dogmatist of geology.

Modified from Adams

Beck, from Adams

LEOPOLD VON BUCH
in field costume.

JAMES HUTTON
founder of modern geology.

Fluids and Exhalations

this effort in Aristotle's theory of earthquakes, but see it much more clearly in Strabo's explanation of floods. Though linked with error, his rains and swollen rivers show that union of natural cause and effect which, after two thousand years, remains geology's great goal.

CHAPTER II

Gropings Succeed Decay

WE HAVE SEEN that the Romans added little to earth science, being content to ignore the planet on whose face they built an empire. That empire grew until A.D. 110 and prospered through eighty additional years. Then it began to decline, a process that gathered impetus till an emperor was captured in battle, Rome itself had to be fortified, and those fortifications repeatedly fell. By 590 the empire had crumbled, the one-time capital had been sacked five times, and a pope threatened by barbarians mourned:

Behold the world in itself is burnt up. . . .
Death is everywhere and mourning, desolation is everywhere and the strokes that smite us and the bitterness that is our daily bread. . . .
Time was when the world held us fast by its delight, now 'tis full of such monstrous blows for us, that of itself it sends us home to God at last.

The classic world had been far from ideal, but we hardly can realize the chaotic degradation of Europe as it sank into the Dark Ages. Here and there crumbling Roman cities survived, but they preserved no traces of Rome's integrating government. In its stead were shadowy kings and warring invad-

Gropings Succeed Decay

ers who murdered, robbed, burned and moved on to diminishing conquests. Medical service and sanitation vanished, letting terrible epidemics sweep from the Bosporus to Britain. Fever stopped Attila's Huns in 452; ten thousand people died in one day at Constantinople; plague ravaged Italy in 565 and 590 and swept through Britain four times between 664 and 683. Famines further reduced stricken populations that clustered into little groups about lords who ruled from newly built castles or old edifices turned into forts. Neither lords nor retainers had time for learning; their need was for walls, swords and a will for vengeance that would warn marauders away.

While Europe sank into ignorance and squalor, classic learning managed to survive in the East. Greek books long since had been taken to Asia, where they were studied in Byzantine schools or translated for rich Syrians of Antioch and Palmyra. By the seventh century Syria, too, was in ruins—an easy conquest for Arabs fired by the promises of Mohammed. Before the year 750 arrived, Moslem faith and arms had built an empire that reached from India across northern Africa and on to the Pyrenees.

Conquering Romans borrowed learning; the Arabs borrowed and then built. Nestorian Christians brought Syrian translations of Greek works to Islam, as well as originals. Jews contributed literature while India added mathematics beyond that of the Greeks. In universities that grew up near great mosques Moslem scholars developed those sources with vigor and ability. They wrote stories and poems, prepared lexicons, created algebra, invented the pendulum, wrote on optics, produced chemical compounds, practiced astronomy, and made discoveries in medicine. They also began to write books on paper, made by methods learned from Chinese prisoners captured in Turkestan.

We look for equal progress in geology, but here the Arabs disappoint us. Even more than the Romans, they were willing to take the earth for granted or get such facts as they wanted from Aristotle, Theophrastus and Strabo. Only the doctor-

The Story of the Great Geologists

philosopher-administrator Ibn-Sina (Avicenna) felt a need to go beyond the classics—and then not very far. In a booklet written at some time during the years 1021–23 he told how banks of soft mud turned into stone, as did water that dripped from cave roofs. Such examples, he said, proved that the ground must contain a "congealing petrifying virtue which converts the liquid into the solid, or earthiness must have become predominant in the same way as in water from which salt is coagulated."

At first this looks quite modern, for geologists now study minerals which settle from underground water and cement loose deposits grain by grain until they become hard stone. But Avicenna's "virtue" was no cement; it was a mystic force which turned other things into stone. Nor did he seem to realize that salt which "coagulated" from the sea had been dissolved in it, just as calcite that forms in caves is merely material that settles from dripping or trickling ground water. Aristotle had erred on both these points, and the great Arab followed him.

On mountains Avicenna followed Strabo, but made one important innovation. Some highlands, he said, were caused by uplift like that which often accompanied earthquakes. But most peaks and steep ridges had been shaped by winds or by streams that eroded valleys in soft, weak rocks but left hard ones as elevations. Yet these high remnants were not untouched, for it "would require a long period of time for all such changes to be accomplished, during which the mountains themselves might be somewhat diminished in size. But that water has been the main cause of these effects is proved by the presence of fossil remains of aquatic and other animals on many mountains." Such remains were left, of course, by eroding waters, but others may have dated from times when strata themselves were formed. At least, said Avicenna, mountain soil contained materials which "perhaps . . . were originally in the sea that once overspread the land."

Arab learning began to enter Christian Europe before the

Gropings Succeed Decay

year 900, at first through importation of books which were turned into poor and erroneous Latin. Still more books were brought by returning crusaders, from whose hands they passed to those of wise men in universities. By 1230 the "Roman" Emperor Frederick II had both Moslem and Jewish philosophers at his court in Catholic and somewhat scandalized Sicily. He also sponsored the introduction of Arabic numerals and algebra, and entertained one Michael Scott, who made new translations of Aristotle's works as annotated by the great scholar Averroes, of Cordova and Morocco.

Thus began the renascence of learning; but not until 1508 did it get around to the earth. In that year Leonardo da Vinci paused during his career as painter, sculptor and engineer long enough to repeat Avicenna's discovery that mountains were produced by erosion of high land. He also announced that rivers were fed by rain and melting snow, that water sank into the ground, where it traveled far through inclined porous strata, and that fossil shells of northern Italy were remains of marine animals which had lived and died when the land was beneath salt water. Once dead, they were covered by river-borne mud that settled upon the sea bottoms, forming layers of rock which in time became part of Europe.

The concept of far-spreading ground water was new; it ruled out the need for primitive fluids or for condensation within the earth. But in his studies of fossils and uplift Leonardo merely rediscovered facts known to Anaximander and Strabo. They interest us now because they had been forgotten —and because a new intellectual climate met their rediscovery.

The ancient world had not censored discoveries in science; with its multiple gods it had no body of revelation which preempted fields of inquiry. Nor, despite the fate of Socrates and Aristotle, did it have any established machinery of orthodoxy. But an established Church found one of these ready-made and rapidly developed the other. In Genesis it received an outline of creation; among the schoolmen and their successors were writers who stood ready to fill that outline with ever-increas-

The Story of the Great Geologists

ing detail. Long before James Ussher set earth's beginning at 9 A.M. on October 26 of the year 4004 B.C., theologians had agreed that our planet could not be much less than five nor much more than six thousand years old. They also accepted the days of creation as literal units, put miracles and magic back into nature, and established the Noachian Flood as earth's greatest catastrophe, universal in its scope.

Holy Writ, however, had not revealed all; gaps were filled from the books of Aristotle, whom the alchemy of faith transformed into Christianity's final authority on science. Works in which he had criticized predecessors and prodded students with speculations came to be studied so slavishly that they discouraged original inquiry and thereby became absurd. "If I had my way," roared gruff Roger Bacon, "I should burn all the books of Aristotle, for the study of them can only . . . produce error and encourage ignorance!" But Bacon's books were prohibited instead, and when Bruno went to Oxford in 1583 he was met by a notice that "Masters and Bachelors who do not follow Aristotle faithfully are liable to a fine of five shillings for every point of divergence." Bruno himself was persecuted for placing Genesis on a level with Greek myths, finally being imprisoned and burned by the Inquisition.

Many a quaint, mold-dotted book shows how dogmas caused timid soul-searching or led to strange speculations. One victim of both was Fallopio, whose chief fame came from studies of man's nerves and reproductive organs. Students apparently showed him fossil shark's teeth, petrified sea shells, bones and teeth of elephants. Fallopio probably knew about Leonardo's ideas; knew, too, that others thought fossils were remains of creatures drowned or buried by the Flood. But Genesis stated clearly that lands and seas were divided on the third day, before animals were created. Leonardo's sea shells therefore violated Scripture, while a deluge that spread from Ararat to Italy seemed to contradict Aristotle, who had said there could be no universal flood. Fallopio concluded that fossils were produced by the action of vapors generated within

Gropings Succeed Decay

rocks. But when those vapors rose with whirling movements they made things which the credulous said were vases and urns buried in ancient times. A huge pile of broken pots in Rome, called the Monte Testaceo, was shaped by those emanations.

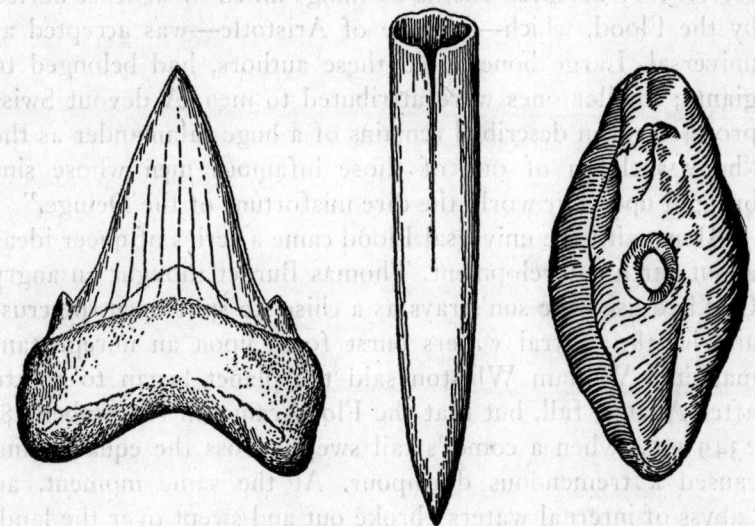

A TONGUE STONE AND TWO CERAUNIAE

At the left, a petrified shark's tooth much like specimens studied by Steno. In the center, a belemnite pen—part of the internal skeleton of a squidlike animal. At the right, a small stone ax said to have fallen from the sky.

Several authors shared this belief; others described prehistoric axes and pestles as "cerauniae" or "thunder wedges" that had fallen from the sky. One that supposedly dropped upon Vienna in 1544 was driven by a flash of lightning which sent it through a house to a depth of twelve ells underground. Another split the oak tree from which it was pried by two farmers who took it to the district tax collector. Such tales reflected an ancient belief, for Pliny (who retold little that was novel) had described cerauniae with the shape of axes which sometimes fell from the heavens.

All this was quite orthodox, for neither vapors nor thunder-

The Story of the Great Geologists

stones conflicted with Holy Scripture. But many naturalists were not content with that; they wanted to prove and elaborate on God's work, much as those of a later generation would seek to multiply proofs of Darwinian selection. Quite sincerely, they accepted fossils as things killed or at least buried by the Flood, which—in spite of Aristotle—was accepted as universal. Large bones, said these authors, had belonged to giants; smaller ones were attributed to men. A devout Swiss professor even described remains of a huge salamander as the "bony skeleton of one of those infamous men whose sins brought upon our world the dire misfortune of the Deluge."

Along with the universal Flood came a series of queer ideas about earth's development. Thomas Burnet thought an angry God had used the sun's rays as a chisel to split open the crust and let the central waters burst forth upon an unrepentant mankind. William Whiston said the planet began to rotate after Adam's fall, but that the Flood came on November 18, 2349 B.C., when a comet's tail swept across the equator and caused a tremendous downpour. At the same moment, an "abyss of internal waters" broke out and swept over the land. From the "chaotic sediment" of this combined flood, stratified rocks settled with great rapidity. John Woodward introduced a variation by making the crust break up and mix with swirling waters and so do its bit to form strata.

These efforts attracted great attention; Charles II was a patron of Burnet's book, while Woodward's became popular in France as well as in Great Britain. But neither added anything to science nor compared with the one geologic work published by a man whose faith soon led him to give up science for activities of the Church.

Nils Steensen, known as Nicolaus Steno, was born at Copenhagen in 1638, of a wealthy family of goldsmiths. He studied medicine at home and in Holland, became a leader in anatomy, and sought a university appointment. But Copenhagen's medical school would not hire him; its jobs went to relatives of the head professor. In disgust Steno traveled to Paris and Flor-

Gropings Succeed Decay

ence, where he became physician to the Grand Duke of Tuscany.

The duke's health was good, his wealth substantial, his disposition generous. Steno found himself with time for study of things beyond anatomy and with money to publish his books. He dissected sharks and studied their teeth, hedging at first but finally deciding that they were identical with petrified "tongue-stones" which (so Pliny and others had said) did not grow upon the ground but fell from heaven during the dark of the moon. Steno also tramped over hills and visited quarries, not so much to collect petrified sharks' teeth as to examine strata. What he saw agreed with his belief that things of earth had been—and still were being—precipitated from fluids. In 1669 he published his conclusions in a volume called the *Prodrome to a Dissertation concerning a Solid Naturally Enclosed within a Solid.*

This now sounds mystical, but the author had no such intent. Steno began with a consideration of shells, "tongue-stones" and other fossils—each obviously a solid enclosed in the solid earth. He felt sure they once had been alive, a fact which proved that the beds containing them were younger than the fifth day of creation, when "God created . . . every living creature that moveth." Some, such as sharks and sea shells, were marine; they meant that the sea once had covered the places in which Steno found them. Other shells had lived in fresh water, and must have been washed about by Noah's Flood "four thousand years, more or less, before our time." Larger bones and teeth were even younger. They belonged to elephants and pack animals of the army which Hannibal led against Rome in 218 B.C.

Steno seemingly knew nothing of granite, nor did he realize the significance of lava. Rocks, he said, were made of stuff that settled from muddy water—salt water if beds contained marine fossils, fresh if they contained fresh-water shells or things that had lived on land. He recognized stream deposits with their coarse, irregular strata, and gave a few simple rules for inferring conditions long past. Thus a thick series of strata, all alike, proved that the water in which they settled remained almost unchanged, but varied strata meant that waters bearing different sediments mingled—perhaps because of windstorms or rains. Other differences were produced where heavy minerals settled before light ones, or where pumice and ashes were sent out by subterranean fires.

Mountains, said Steno, were not like animals and did not really grow. Some, as everyone knew, were volcanoes which belched out ashes and hot stones from fires burning deep underground. Other mountains had been forced upward by rising exhalations, or were tossed into jumbles of bent, broken strata when subterranean caverns collapsed. Still others had been shaped by running water, just as cliffs and gorges were still being cut by streams in foothills of the Alps. Cracks produced by uplift or collapse made "receptacles" for minerals, some of which were produced by vapors and some by other means.

Steno applied his general theories in a geologic history of Tuscany, the first one of its kind ever written. The record had six phases, or epochs, which he thought might be traced over the entire earth. First came a time of marine submergence during which rocks accumulated without a trace of life. Then land appeared as a plain raised above the sea. During the third epoch, the plain was broken into mountains and hills after cavities had been "eaten out by the force of fires or waters." Then the sea advanced again, filling valleys with sandy sediments which contained great numbers of fossils. Land reappeared in the fifth epoch, with rivers whose mud built up deltas and plains that each year spread farther and farther into the retreating sea. During the sixth and final epoch, plains were uplifted and eroded while subterranean fires destroyed strata, allowing those at the surface to break and so form new precipices and hills.

The *Prodrome* was an outline, a great beginning, but Steno went no further. A devout Lutheran, he now turned Catholic

Gropings Succeed Decay

—some say through the influence of a nun; some say through that of a friend in Paris. He lost interest in geology and medicine; in time he took orders, was made a bishop, and became chief organizer of propaganda in northern Germany. There he lived a life of extreme and unbalanced asceticism which so undermined his health that he died in 1686.

Steno's genius did not endure, nor was his work accepted by others. Martin Lister, for instance, read the *Prodrome* and then gravely informed the Royal Society of London that "for our English inland quarries I am apt to think that there is no such matter as petrifying shells in the business: but that these cockle-like stones were never . . . any part of any animal."

Edward Lhuyd, of Oxford, speculated upon spore-bearing vapors that arose from the sea, rained down into rocks and there produced the stony "marine bodies which have so much excited our imagination." The "Roman gentleman" Colonna announced that mountains grew upon earth like trees, though more slowly; perhaps only a hundred paces in a thousand years. Abbé Anton-Lazzaro Moro disagreed, saying that mountains, islands and even stratified rocks were products of eruptions.

Abbé Moro (1687–1740) deserves our attention, not because he learned anything worth while, but because he sums up the weaknesses of his age as well as of ages before it. He began, as Aristotle urged, by collecting data, most of which came from books rather than from observation. Having assembled records of some sixteen islands suddenly built by eruptions, he announced that nature always produced similar results in one, and only one way. This meant that *all* islands and their mountains were volcanic: Sicily, Borneo and Britain as well as three Greek islets in the Gulf of Santorin, each piled up by a single eruption on the floor of the Aegean Sea. And since island mountains could not differ from those of mainlands, it followed that all ranges and peaks were products of vulcanism.

As for stratified rocks; their origin was clearly revealed by

The Story of the Great Geologists

the materials of which they were made. Moro looked at shales, sandstones, conglomerates; in them he saw only ashes, cinders or worn bits of stone that had come from volcanoes. They must, he said, have been blown from craters, after which they fell upon land or water where they accumulated in layers. Limestones began as calcareous masses which had been heated into soft, doughy blobs that ran together as they fell. If one watched he could see just such masses emerge from the crater of Vesuvius.

Whence came so much heat? From subterranean fires that raged through the whole center of the earth and filled spaces from which lava had been ejected. Moro saw no need to specify fuel, for in those days fire was a thing in itself, independent of wood or coal. But its origin was another matter: one to be settled by revelation. "It pleased the Creator of all things," wrote Moro, "that great subterranean fires should be kindled on the third day, when dry land was to appear according to the sacred account in Genesis."

In other words, God himself set the earth's interior ablaze and so by outpourings or upwellings of lava produced the first lands to rise from the primeval ocean. These lands then cooled with such rapidity that they could bring forth "grass, and herb yielding seed . . . and the tree yielding fruit" before evening closed the third day of creation. And since the one inundation allowed by Genesis came from rain, it was foolish to talk of marine creatures whose shells and petrified teeth lay in mountains.

Poor, unbalanced Steno had done so, but both Steno and his *Prodrome* were dead.

CHAPTER III

Maps and Ancient Volcanoes

Paved roads climb hills and dip into valleys that lead to the upper Mississippi or wind across central New York. From these valleys one looks upward to cliffs where limestones, sandstones and shales rise in gently tilted strata that lie one upon another like the leaves of a book. They form series that run for hundreds of miles, appearing and reappearing in heights that overlook black-soiled lowlands where no trace of solid rock can be seen.

In Abbé Moro's time those gradient strata were known only to indifferent Indians and a few white pioneers. Europe's rock series, too, were almost unsuspected; not because they lay in wild, unsettled regions but because no one had devised a method for recognizing different formations and tracing them from place to place. That task was begun by a great though crotchety Frenchman who made his first reputation as a collector of plants.

Jean Etienne Guettard was born in the autumn of 1715, at a village thirty miles south of Paris. His grandfather was an apothecary—in those days a combination of druggist, lay doctor, public councilor and amateur naturalist. He let little Jean

The Story of the Great Geologists

play in his shop and took the growing boy on walks which modern jargon would call nature hikes or field trips. Jean had a keen eye for plants; soon he was finding species which *Grand-père* could not identify. In time his finds brought praise from the great Jussieu, of the botanical garden in Paris. It was agreed that Jean should have a career in science, and since the way to do that was to become a doctor, he took a degree in medicine. The degree seemed—and was—a smaller achievement than membership in the Royal Academy of Sciences at the age of nineteen.

Those were days when wealthy men often hired scientists, yet accepted them as peers and friends. Instead of dispensing pills and treating patients for gout, young Dr. Guettard "entered the household" of Louis, Duke of Orléans. This son of a dissolute politician and minister was a pious, retiring amateur in science who also spent much time translating the Psalms. Guettard took charge of his collections, traveled with him, and was sent on independent journeys. When the duke died his soldier son gave Guettard a small pension and an apartment in the family palace at Paris.

Most scientists select their field and stick to it; but botany, his first love, led Guettard to a change. On trips through France and western Germany he noticed that certain kinds of plants were found only where the ground contained special minerals or rocks. To make sure he took samples, wrote down notes, and then made new journeys to check and recheck his data. Soon he was so busy with minerals and rocks that he gave little thought to plants apart from the ground beneath them.

Friends in the Academy of Sciences shook their heads: where was all this labor leading? They learned when Guettard advanced the conclusion that rocks and their minerals were not scattered hit-or-miss but were arranged in bands that ran through mile after mile of country. Once the width and course of a band were known, it forecast both the minerals and plants to be found where it passed through unfamiliar regions. Such foreknowledge seemed amazing. Would not *Monsieur le doc-*

Maps and Ancient Volcanoes

teur make further observations, to be embodied in a memoir and map for publication under auspices of the King?

There is small chance that the weak and licentious Louis XV knew or cared about Guettard's work; the phrase *par ordre du roi* was only a pleasant fiction. But there was nothing fictitious about the eighteen hundred miles which Guettard traveled on foot, on horseback, and in coaches to get still more specimens and notes. The result was a quarto memoir with two maps which the Academy published in 1752. On the larger map, symbols showed where various rocks and minerals had been found, while shading marked two bands of seemingly related formations that surrounded an irregular oval with Paris near the center. That oval was called the "sandy band"; a series in which sandstones, millstones and limestones, as well as gunflints, were found. Next came the "marly band," where most of the rocks were hardened marls and fossil shells were common. The outermost, or "schistose," band contained useful ores in addition to coal, sulphur, bitumen, slate, marble and granite. It was, in short, the chief source of French wealth in minerals.

Drawing these oval bands on his map, Guettard saw that they were cut by the English Channel and Straits of Dover. Eagerly he ransacked two books, Childrey's *Britannia Baconica* and *Ireland's Naturall Historie,* by Boate. Though old and none too accurate, they enabled him to extend the French bands into England and so establish his main thesis. Rocks did lie in some definite order, and each group carried its "mineral products," no matter where it was found. Guettard therefore made his second map, which showed the distribution of minerals and rocks from North Ireland to Spain and the Mediterranean coast.

While seeking minerals, Guettard discovered fossils that ranged from castoff remains of ancient sea dwellers to bones and teeth of elephants. Near Paris he found great numbers of shells; in Normandy he dug out petrified sponges and was laughed at by farmers who said they were apples and pears

The Story of the Great Geologists

which had turned into stone after dropping as windfalls. From Angers came trilobites; Guettard compared them to crabs and prawns and gave an excellent picture of *Illaenus* with its bulbous tail and head. To answer those who still called such things mere freaks of nature he prepared a handsomely illustrated treatise *On the Accidents That Have Befallen Fossil Shells Compared to Those Which Are Found to Happen to Shells Now Living in the Sea.* He showed ancient oysters attached to other shells, snails covered with barnacles and worm tubes, shells in which worms and sponges had burrowed. Other shells had been wave-worn or broken just as shells now are eroded, chipped and battered to pieces while they roll to and fro in the sea.

As a child Guettard had played in the shade of a sandstone pinnacle known as the Rock of the Good Virgin because it bore some resemblance to a woman holding a child. Coming back as a man, he found the rock crumbling; then it fell into pieces which were washed into the little valley below. Other rocks meanwhile became prominent—not because they grew, as villagers said, but because detritus that once masked them had been carried away by rivulets that rushed downhill during rains. Farther on, the rain water had dug gullies where Jean and other children once played on smooth, grassy slopes.

Such changes had been going on for ages but no one had thought much about them or even noticed their cause. Guettard made them the text of a work on "degradation"—we should call it erosion, perhaps with emphasis on soil. He showed how running water wore land away; how swift streams dug box canyons; how subterranean water dissolved rocks to make caves or caused landslides by turning clay into mud. He also described the action of waves, which cut cliffs as they battered against hilly shores. Fragments worn from those shores then sank to the bottom, making new beds of rock, just as fragments carried downstream by rivers formed layers upon flood plains and deltas.

All this now seems commonplace to us, who know the men-

CHARLES LYELL as an elderly man.

After Mrs. C. Lyell

Phillips

WILLIAM SMITH
who first put order into the succession of strata.

Part of William Smith's great map, from Maclure's copy in the
Academy of Natural Sciences, Philadelphia.

Maps and Ancient Volcanoes

ace of deepening gullies and the terrible force of waves that roll landward from the open ocean. But in contrast to Moro's volcanic strata, Guettard's statements were revolutionary. So was his discovery of volcanoes in the Auvergne district of south central France.

That discovery was made in 1751, as Guettard and his friend Malesherbes rolled in a creaking coach toward the cathedral town of Moulins. Malesherbes, a great lawyer, was looking for plants; Guettard had an eye out for minerals to be recorded on his map. Now and again he jotted down a symbol —and then stared when he saw mileposts cut from black, porous stone that matched printed descriptions of lava. Whence came these mileposts? The coachman couldn't tell, but in Moulins people who should know said they had been brought from Volvic. Quarries there produced the rock for sale in the country round about.

Guettard and Malesherbes hurried on, confident of their goal as the black stone grew more and more common. Riom's gloomy old houses were built of it; townspeople told how to reach the quarries, then less than six miles away. Soon the travelers could distinguish dark pits in the side of a hardened lava stream that reached downward from a granite ridge and overlay the plain below it. Above the ridge rose a cone-shaped hill, or *puy*, with a crater and a gap at one side from which the lava had flowed.

Guettard had never seen a volcano, but he knew those of Italy from books. Climbing the rough slopes with Malesherbes, he recognized pumice tossed out during eruptions and porous, glassy scoria that had hardened as steam and other gases bubbled to the surface. Then the two men scrambled over solid lava, tracing its sheetlike flows as well as partings of clay or old soil which marked periods of quiet and erosion that had passed between eruptions. Later, with a Clermont apothecary as guide, a trip was made to Puy de Dôme, whose crest gave a view of fifteen or sixteen volcanic cones as well as many lava streams and fields of pumiceous cinders.

The Story of the Great Geologists

Though grassy, the cones looked so fresh that they seemed to threaten new eruptions almost any day.

Coaches, guides, lodgings—all cost money, and Guettard had little to spend. Besides, Malesherbes was a busy judge and censor who had work to do in Paris. There the two men returned, and on May 10, 1752, Guettard described their discoveries before the Academy. His first account grew into a memoir that was published in 1756.

That memoir contained brand-new information, but it also reflected its time. The modern geologist who discovers a new volcanic field describes its cones, flows, cinder fields, and perhaps the characters of its lavas. Guettard did all that very well; in spite of dull, pedantic prose he shows us the steep-sided hills with their circular craters, the porous pebbles of pumice, and the ropy ridges that had formed as cooling, viscous rock ran downhill. But custom demanded an explanation, and here Guettard relied on a theory whose only virtue was its age. "For the production of volcanoes," he wrote, "it is sufficient that there should be within these mountains materials that can burn, such as petroleum, coal or bitumen, and that from some cause these things should catch fire. The mountain thereupon will become a furnace, and the fire that rages inside it will be able to melt and vitrify the most resistant substances." Hot springs near the *puys* of Auvergne were warmed by such fires, which either had not burned out or were preparing to burst through the ground in new and violent eruptions.

We now know that all this was nonsense; that volcanoes do not get heat from fuel and have little or no true fire. But this error of Guettard's was not too important, nor did it then matter much that he mistook the columnar type of lava called basalt for a sedimentary rock. What counted was the fact that he did recognize ancient volcanoes, tracing their history in cinder layers, old soils, and solidified lava flows. More important still were his "mineralogic" maps, whose shaded bands were the predecessors of modern geologic formations, and whose data demonstrated the need for thoroughgoing field

Maps and Ancient Volcanoes

work. His studies of fossils and erosion also showed the worth of detailed firsthand study, a contribution which then was quite as important as their actual additions to knowledge.

Guettard was no man to hoard his data or to spend a lifetime on one or two reports. He published some two hundred articles of varied lengths, as well as a half-dozen massive books with abundant and detailed plates. Those published by the Academy used up some of the royal funds which even the industrious Louis could not squander on luxury and courtesans.

Still, we must not picture Guettard as a man who rushed into print with inadequate or hasty papers. He took six years to prepare his two mineralogic maps, with the memoir that explained them. Five years were devoted to a work on the strata and fossils of Angers, and comparable periods on other works. With no classes to teach, no wages to earn, there was time to work thoroughly and still get a great deal done.

Nor did Guettard write to gain influence by advertising his own achievements. His pension met his bachelor needs; since he wanted no official job, he need neither appear better than he was nor please men of influence and position. Instead, he developed a crusty manner which often angered them, causing minor scenes in Academy meetings and quarrels with such good friends as Malesherbes. Thanked for a favorable vote, Guettard once grumbled, "You owe me nothing . . . because I don't like you!" and curtly turned away. He criticized poor papers mercilessly, and when eulogies of dead members were read he would march up to the secretary and announce: "You are going to repeat a lot of lies. When my turn comes I shall want only the truth told about me!"

The crusty critic's turn approached as he neared seventy. First he began to suffer from periods of abnormally deep sleep, during one of which his foot slipped so close to an open fire that it was badly burned. Guettard did not complain, but repeatedly assured his doctor that this or that remedy was useless, since it could not ward off an impending "stroke."

The Story of the Great Geologists

Soon the prospect of death so seized his mind that he refused to make long visits or even dine out with friends. Not that he feared the effects of such trips. He merely was unwilling to bother an acquaintance by being stricken in his home!

In spite of this, Guettard walked to every meeting of the Academy of Sciences, with no other precaution than a card bearing his name and address which was tucked in an outside pocket. This information would get his body home if he collapsed. Why spend money on coach fares to delay that inevitable stroke?

Guettard died in January 1786, at the age of seventy-two. He never went back to the volcanoes; at least, not for a detailed study that would amplify the memoir based on his visit with Malesherbes. He was too busy with new mineral maps, fossils, erosion, and the river basins of France. Volcanoes he would leave to other investigators.

Chief of these was Nicolas Desmarest, an inspector of industries for the Crown. Born into a poverty-stricken home in 1725, Nicolas could barely read or write when his father died fifteen years later. The lad was taken in by Oratorian priests who gave him a free education at their colleges in Troyes and Paris. Becoming a teacher, he entered science with an essay on the land bridge that once linked France with Britain where now lies the English Channel. His work was more thorough than that of Guettard, whose interest was not in land bridges but in continuous belts of rock. The essay brought friendships with the mathematician, D'Alembert, the Duc de la Rochefoucauld and other important men, and led to an appointment as inspector of industries in 1757.

This inspectorship was an uncertain affair, with low pay, varied tasks, and long periods of leisure which Desmarest used for field work in French geology. He traveled alone and on foot, with dry bread and cheese as his principal foods. He avoided manor houses where he might have been a guest, as well as ordinary inns in which he would have paid for lodgings. He preferred to sleep without honor or cost on a bench in

Maps and Ancient Volcanoes

some farmer's kitchen or on the earth floor of some herdsman's hut.

In 1763 Desmarest got around to visit the volcanoes and lava fields near Volvic. Walking from Clermont to the Puy de

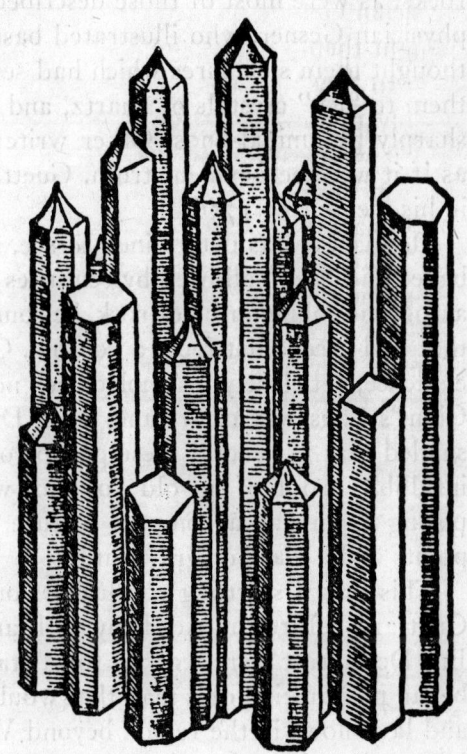

BASALT COLUMNS

As figured by Conrad Gesner in his *De Rerum Fossilium,* published in 1565. Note endings which resemble those of quartz crystals.

Dôme, he climbed a plateau where granite was overlain by coarse, bubbly cinders and old, burnt soil, with a capping of black rock composed of columns that stood on end like the logs in a pioneer stockade. Each column was pressed against its neighbors, most of which had six flat sides that met in obtuse angles. *"Parbleu,"* Desmarest exclaimed to himself. "this must be basalt!"

Basalt was neither rare nor new; it was a rock found throughout the world, from Scotland to the Celebes and the

[35]

deserts of what then was New Spain. Pliny had described and named it; *Stadtarzt* Georg Bauer (called Agricola) had revived Pliny's name in a book *On the Nature of Fossils* that was published in 1546. Bauer's "fossils" were minerals and rocks, as were most of those described by the Swiss naturalist-physician Gesner, who illustrated basalt columns in 1565. He thought them structures which had settled in water, compared them to "icy" crystals of quartz, and even gave some of them sharply pyramidal ends. Other writers reiterated his opinion as if it were self-evident truth. Guettard himself had done so in his memoir of 1756.

Desmarest was determined to see, not believe, even though belief had been hallowed by centuries of repetition. Tramping across the columnar black rock, he found it to be a sheet of lava that had spread out from a volcano. Other sheets gave similar evidence, yet closely resembled such now famous basalts as the Giant's Causeway of Ireland. Soon Desmarest was "fully persuaded that in general these groups of polygonal columns are infallible proof of an old volcano, wherever the stone composing them has a compact texture, spangled with brilliant points and a black or gray tint."

This was a startling conclusion; one that would have sent Guettard before the Academy with an article for publication. But Desmarest merely asked for a state surveyor and in 1764 began preparation of a map that would show all the volcanoes and lava flows in the region beyond Volvic. With this map he returned again and again, never willing to call it complete or send it to an engraver. Better let his ideas go unaccepted than put forth an unfinished work!

Desmarest showed equal patience and caution when he entered other fields. Having discussed the nature of French volcanoes and their lavas, he proceeded to deal with processes of erosion by which they were destroyed. This led to a study of volcanic periods and related changes in earth's surface and brought what then was the novel conclusion that streams eroded their own valleys instead of flowing through gorges or

Maps and Ancient Volcanoes

lowlands which they found ready-made. Desmarest sketched his ideas in 1775 but waited thirty-one years to publish them. Meanwhile other men reached the same conclusion, and quite reasonably received credit while his own work was ignored. The cautious inspector did not complain, nor did he seem to realize that by holding back he was making fellow scientists waste time.

Indeed, Desmarest seemed to care very little for credit, for his fellow scientists, or for affairs of the world. He was pleasant, kindly, dignified; a man who became almost majestic as decade followed decade. But he read no novels, saw no plays, cared little for the arts; in Rome museum curators actually feared that he might chip bits from statues to determine the stone of which they were made. At a party he interrupted a thrilling tale of shipwreck to ask primly whether the vessel had foundered on basalt or limestone. He refused to change the cut of his clothes and in 1815 wore a wig that had not been too fashionable seventy years before. Though a family man, he spent every Sunday with a friend to whose village he went on foot. The friend died, but Desmarest kept up his Sunday trips even when, at eighty-five, he had to give up walking and bounce over rough roads in a coach.

His kindness did not extend to Guettard, for whose one hurried study of volcanoes he voiced vitriolic contempt. "Can we regard as a true discovery," he snarled, "the simple recognition of the products of volcanic action, when the facts are presented with so little order and so much confusion? Such a discovery implies a reasoned analysis of all the operations of fire . . . so as to reveal the ancient conditions of all the volcanic regions. Without this it is impossible to dignify the recognition of a few stones with the name of a discovery that will advance the natural history of our earth!"

To this outburst no reply was given, for Guettard himself had been dead nine years and his most faithful friends had fallen under the guillotine. Nor did Desmarest himself reasonably explain subterranean "fire" or comprehend its opera-

The Story of the Great Geologists

tions. We can excuse his attack only on the ground that he fell victim to one of those unreasoning antipathies that appear among scientists as well as back-yard gossips. It may have been aggravated by the fact that both he and the man whom he belittled were very much alike. Both cared little for the niceties of life; both were persistent investigators who did what they felt should be done without thought of personal advantage. Both erred and both—despite Desmarest's judgment—made significant discoveries. Even more important was their proof that such discoveries could come from careful field study rather than from speculation. That fact might and would be ignored, but it never could be so far forgotten as to put geology back to the uncontrolled suppositions of Burnet or Abbé Moro.

CHAPTER IV

Geology by Dictum

THE SAXON CITY of Freiberg lies on a bleak northern slope of the Erzgebirge, whose name signifies Ore Mountains. Near-by mines reach veins rich in silver, lead, copper and other metals, with crumbling tunnels and tree-grown dumps that date back to medieval times. Both smelters and foundries belch out smoke, smudging the windows of cotton mills in another part of the city.

Here in 1765 was founded a school of mines (*Bergakademie*) whose director, a pioneer mineralogist, took a fatherly interest in really good students. One of these was a youth named Werner, who already knew more about rocks and ores than did most professors or miners.

Abraham Gottlob Werner was born on September 25, 1749, in the smelting town of Wehrau. His father was an inspector of iron works; his mother gave him stones to play with almost before he could talk. At the age when modern children use blocks and wooden mallets, little Abraham stood bits of sandstone on edge and split them with a hammer. As a reward for reciting *ah, bay, tsay* and on to *tset*, his father let him look at minerals and told him what they were good for. Later the boy pored over the close-printed pages of mining lexicons, collected specimens from dumps, and begged his willing father to name them. At the age of fifteen he took his first job, helping Herr Werner manage blast furnaces at Wehrau.

But the son was not content to remain a provincial smelter official. In 1769 he enrolled at the academy in Freiberg, making visits to mines and mills in periods between classes. He also took longer trips through the mountains to examine rocks, check ore deposits, and get minerals for his collection. This was the envy of fellow students long before young Werner went on to Leipzig for three years of general study at the university. As a student he published an article *On the External Characteristics of Fossils,* by which he meant minerals. In 1775, when only twenty-five, he returned to Freiberg as inspector of mines and teacher of mining and mineralogy at a wage of three hundred thalers per year.

Seldom has a school paid so little for so much. The youthful inspector began to lecture on minerals, mining and ore dressing on the plan of a textbook published while he was in Leipzig. Facts, however, were largely his own; instead of parroting data which he had read, he told of things seen in mines and mills throughout the Erzgebirge. Soon he added discussions of machinery and tunnels, mining finance, history, mineral law—in fact, on almost any subject that appealed to him and would profit his listeners. On one day a mineral specimen might lead to a talk about ore bodies; on another it would provoke a discussion of engineering problems to be solved when those ores were mined. Blocks of stone illustrated lessons on rocks, but they also were texts for long digressions upon the geography of Europe. In sonorous words the lecturer described granite mountains, valleys floored by shale, and contrasting regions underlain by limestone or quartzite. If students asked what kind of people dwelt in those regions, Werner was ready with an account of who they were, what they did, and how their whole lives were determined by differences in rocks.

Such teaching was not orthodox, but it was stimulating.

Geology by Dictum

Students praised this remarkable man who never wrote a lecture in advance and was ready to lead his students through almost any field. Others dwelt upon his personal charm—his ringing voice and commanding language, his ready smile and enthusiasm, his trick of brightening dull topics with a shy yet heavy-footed humor. Still others praised his vast learning and the thoroughness of his teaching. For there were no loose ends in Werner's courses; each subject had its proper place and was touched upon even when little could be said about it. Demonstrations were given as well as lectures; large classes were broken up into groups so that everyone could see and hear. With friendly smiles and keen questions Werner made sure that each group mastered one subject before going on to the next. Out of doors he questioned, commented and explained as he led his classes through Saxon valleys or into dripping mines. "Let us go and see!" he was fond of saying. "We must know rocks as they are down in the mines, not as specimens in the *Bergakademie's* collections!"

As the tale of such teaching spread it brought more and still more students over the rutted roads to Freiberg. Germans came from other kingdoms, to be followed by Scots, Frenchmen, Italians and Spaniards. From raw youngsters to important officials, they showed that what once was a local school had become Europe's great center for training in geology.

Not that Werner employed this term, which had been introduced in 1735 by an English author. To the inspired inspector in primly curled peruque, geology was literally *earth discourse*—a pretentious structure of speculation about the origin and early history of our planet. Such nonsense was anathema to Werner, whose goals were order and good, solid fact without trace of theory or guesswork. To show this he borrowed another term, geognosy, which signified *earth knowledge*.

This insistence on fact was valuable, as was Werner's love of order. When he began to teach in 1775 the science then called mineralogy was a conglomeration of truth, error and traditional fiction with no better organization than that of the alphabet. Werner had to weed out much worthless clutter, put order into what remained, and invent a vocabulary which would say what he had to say with some degree of precision. Almost inevitably his system was artificial; being German, his descriptive terms were massive polysyllables which can only be translated by such phrases as "bent like a fortification," "indeterminately curved-angular" and "not especially difficultly breakable." They were hard to pronounce and often dubious in meaning, yet even critics who found them inadequate agreed that they were better than nothing.

From minerals Werner turned to the earth, its rocks, and its history, still insistent upon definite knowledge and scornful of speculation. This time, however, he did not invent; he borrowed the ideas of Herr Johann Lehmann, who had taught mineralogy and mining at Berlin before going to Russia in 1761. There he became professor of chemistry and director of the Imperial Museum, serving in both capacities till death came when a retort full of hot arsenic blew up in his face.

Lehmann thought that the earth had begun as a vast blob, or "void," of muddy stuff uniformly dispersed through water. At creation's first stroke the mud settled, forming a shell whose high parts became continents and islands as water drained to an abyss within. Much later than this came Noah's Deluge, a universal flood that washed sediment from primitive mountains and deposited it on their slopes. After that other formations appeared, some as volcanoes or lava flows and some as stratified deposits laid down by still later floods.

For years Werner developed Lehmann's theory, adding details, lopping off some excesses, and providing new ones of his own. He thus recast the whole into a system of geognosy which seemed to meet his desire for an earth science to be taught as incontrovertible fact. He so outlined it in a quarto booklet published in 1787.

We may doubt whether twenty-eight pages ever covered a greater field, displayed more obvious errors, or contained more self-contradictions. Here was a man of thirty-eight who

Geology by Dictum

had neither read widely nor traveled far beyond Leipzig, yet who undertook to tell earth's story from the very moment of creation and for every continent. That story would be strictly factual, yet to write it Inspector Werner must assume, guess and speculate. He began by assuming that Lehmann's earth ball had really existed and that its solids were precipitated while liquid disappeared. He guessed at the nature of both minerals and rocks, and he speculated recklessly by arranging those of Saxony in formations typical of the world.

These universal formations numbered four, all found within the vicinity of Freiberg. First and oldest was the Primitive, or Primary, a thick series of precipitates that settled during those early times before dry land appeared. It included granites, slates, basalts and some marbles, as well as layered deposits called schist, which often had solidified in beds that arched or stood on end. None of these was more recent than creation and none, said Werner, contained fossils. By this he probably meant—quite erroneously—that they lacked substantial nuggets of metal or large and angular crystals. For Werner seldom if ever used the word "fossil" for remains of animals or plants.

Primitive rocks formed the cores of great highlands, where they had settled and hardened while the sea was amazingly deep. On their slopes and beyond lay the Flötz Gebirge; literally, stratified formations or "mountains." These included limestones, sandstones, conglomerates and the like, as well as chalk, gypsum, shale and coal. Part of their material also was precipitated, but coarse stuff consisted of fragments worn from the Primitive highlands. "Fossils" were found in many beds, but Werner did not tell what they were.

The Flötz was followed by ashes and lavas of the Vulkanische, or volcanic, Gebirge, which also included strata changed by intense heat. Such strata, however, were not extensive, nor did lavas and volcanoes account for significant parts of the earth. The latter, in fact, were superficial affairs heated by coal beds of the Flötz which somehow burst into

The Story of the Great Geologists

flames underground. "It was only after the deposition of the immense repositories of inflammable matter in the Flötz-trap," wrote one of Werner's followers, "that volcanoes could take place; they therefore are to be considered as new occurrences in the history of nature. The volcanic state appears to be foreign to the earth."

Above and after the Flötz came "washed" deposits of gravel, sand, clay and soil which now cover major portions of the land. We should call many of them alluvial, but hesitate to say just what meaning lay in the adjective *aufgeschwemmte* as it rolled from Werner's tongue. It probably had much more to do with waves on the shrinking original sea than with rivers which spread worn particles of rock across deltas or flood plains.

One would like to think that this "system" was tentative; something to be improved upon or discarded as geognostic science progressed. Actually, however, Werner showed no such caution; sure that his work was based upon fact, he both wrote and spoke with assurance that brooked neither contradiction nor doubt. Four years later, although he had been forced to move columnar basalts into the Flötz and was preparing to insert a Transition Formation above the Primitive, he could detail his assurance:

> In summarizing the state of our present knowledge, it is obvious that we know with certainty that the Flötz and Primitive formations have been produced by a series of precipitations and depositions formed in succession from water which covered the globe. We are also certain that the fossils which constitute the beds and strata of mountains were dissolved in this universal water and were precipitated from it; consequently the metals and minerals found in Primitive rocks, and in the beds of Flötz deposits, were also contained in this universal solvent, and were formed from it by precipitation. We are still further certain that at different periods different fossils were formed—at one time earthy things, at another metallic minerals, at a third time some other fossils. We know, too, from the position of these fossils one above another, how to determine with the utmost precision which are the oldest and which the newest pre-

Geology by Dictum

cipitates. We are also convinced that the solid mass of our globe has been produced by a series of precipitations formed in succession (in the humid way); that the pressure of the materials, thus accumulated, was not the same throughout the whole; and that this difference of pressure and several other concurring causes have produced rents in the substance of the earth, chiefly in the most elevated parts of its surface. We are also persuaded that the precipitates taking form from the universal water must have entered into the open fissures which the water covered. We know, moreover, for certain, that veins bear all the marks of fissures formed at different times; and, by the causes which have been assigned for their formation, that the mass of veins is absolutely of the same nature as the beds and strata of ordinary formations, and that the nature of the masses differs only according to the locality of the cavity in which they occur. In fact, the solution contained in its great reservoir (that excavation which held the universal water) was necessarily subjected to a variety of motion, whilst that part of it which was confined to the fissures was undisturbed, and deposited in a state of tranquillity its precipitate.

We know . . . we are certain . . . we are convinced; each radiated assurance, yet each introduced a speculation whose only basis was error. Werner rejected Desmarest, Guettard, Steno, Leonardo; he offered a theory of mineral veins which was even further from truth than Aristotle's exhalations. He told students to believe in a primitive planet *covered* with water, yet whose solid core came into existence by precipitation from the primeval sea. Ignoring Avicenna as well as Desmarest, he went on to describe great floods and other "powerful influences" which had shaped cliffs, had dug narrow gorges, and had worn such valleys as those near Wehrau and Freiberg.

The great man spoke; his students believed and went forth to make the earth yield illustrations of the Wernerian system. In the flowery words of one biographer, they worked in the "cabinet," or museum, as well as "at the summit of the Cordilleras, in the midst of the flames of Vesuvius and of Etna, in the deserts of Siberia, in the depths of the mines of Saxony, of Hungary, of Mexico, of Potosí." In all these places they found facts to support Werner's system or managed to make contradictions seem to agree with it.

For discordant data did appear and had to be explained away. Thus young granites and porphyries were found; since they could not be Primitive, these discoveries were squeezed into the Flötz as late precipitates which somehow mimicked those of very ancient times. Great faults in the Alps and Andes were minimized; made small, they then could be regarded as cracks produced by varying pressures at or near the earth's surface. Pumice was found to be plentiful; it therefore ceased to be a volcanic rock, since eruptions were known to be so unimportant that their products were almost "foreign to the earth." When a Scotch geognost could find no way to drain the deep Primitive ocean he declared that the problem need not be solved, because Wernerians were "fully convinced" that such an ocean had existed, regardless of its possible fate. Werner himself supposed that much of the early earth's atmosphere and water had been captured by some other celestial body. Yet he was not disturbed when followers brought part of the water back to deposit basalt upon newly built Flötz mountains of sandstone, limestone and shale.

Such were the ideas, the intellectual evasions, which were to free earth science from speculation by replacing tenuous theories with facts. Werner elaborated his system through twenty-six years of informal lectures, yet never wrote it down in detail. Since the world asked for a treatise he promised it, but published only a few short articles and small volumes on special subjects. Yet he grumbled when other men brought out books and waxed wrathful when enterprising ex-students planned to issue reports of his lectures. Such nonsense, he warned, must cease at once; the proposed reports were unauthorized, obsolete, and probably full of mistakes. Possible purchasers should save their minds and their money for a series of works which even then—in 1791—were being revised for publication. Containing authoritative statements of Werner's system, these books would "appear forthwith, one after

Geology by Dictum

another, enriched by his latest observations and discoveries." Actually, the books seem never to have progressed beyond a table of contents which was printed in 1783. Held twenty-eight years, it finally was released in 1811, perhaps as another warning to profit-minded intruders.

Werner doubtless intended to produce those books, but never got around to doing so. Admirers hinted that he was too busy; that his devotion to classes and special students left no time for composition. A more likely cause was an aversion for writing which developed through the years until it became an obsession. Most teachers of his time put down every sentence of their lectures on paper, but Werner could not bring himself to do more than scribble a few key words on slips. He sent no letters and answered none; when friends complained about this he felt so hurt that he ceased to open mail. An author who had sent him a manuscript could not get it back, and after long search it was found among hundreds of untouched parcels in one of Werner's rooms. At another time he kept a messenger waiting two months—not for a book or even a letter, but merely for his signature on a family document.

Such isolation made for a placid life, into which students brought only adulation and critics were not admitted. They could not even come by way of print, for when hostile articles began to appear Werner stopped reading journals. His loss, perhaps, was not very great, for by that time his system was so thoroughly perfected that it had room only for corroborative details.

Yet placidity could not hide one fact: that Inspector Werner was growing unwell. An ordinary lecture left him weary; when he really let himself go the result was nervous exhaustion preceded by such an outpouring of sweat that he must quickly change his clothes. He dare not stand or sit in a draft and bundled himself up in rooms which to others felt unpleasantly stuffy. Yet those were not times when people were readily disturbed by stale air!

Into this carefully nurtured seclusion burst Napoleon's

The Story of the Great Geologists

army of occupation. Its mere presence annoyed Werner, who complained when soldiers were arrogant or brooded because fellow townsmen had to feed the invaders, house them, and provide fodder for their horses. Melancholy so aggravated his obscure disorders that he finally became ill. Unable to teach or even care for himself, he left Freiberg for Dresden and the home of a sister, who nursed him. There he died on June 30, 1817, long after the offending Napoleon had been taken to exile and Europe had returned to peace.

CHAPTER V

Those Skeptical Scots

WERNER'S COLLAPSE staggered the *Bergakademie,* where enrollments dropped at a rate unknown during the wars. What was worse, a committee appointed to deal with the problem reported that nothing could be done. There were heretics in the field, it appeared; men who not only disagreed with Werner, but who taught a system of earth science that professed to be much closer to truth than geognosy. Once impotent, these critics now were able to attract the students who during the inspector's life would have come to Freiberg.

Rebellion, indeed, had been growing through more than twenty years. As long ago as 1796 a Scotch author named James Hutton had attacked Werner's essential theory of creation, along with the notions of Burnet, Woodward, Lhuyd and other speculators. It was natural, Hutton admitted, to seek out origins; man always wants to know the beginning of things to be found around him. "But," he objected, "when a geologist shall indulge his fancy in framing, without evidence, that which had preceded the present order of things, then he either misleads himself, or writes a fable for the amusement of his reader. A theory of the earth . . . can have no retrospect to

The Story of the Great Geologists

that which had preceded the present order of the world; for this order alone is what we have to reason upon; and to reason without data is nothing but delusion."

The author of this forthright criticism was a man who cherished strong likes as well as dislikes, who had jovial yet reserved habits, and who worked with untiring industry. After a career as doctor, scientific farmer and manufacturing chemist, he had established himself in Edinburgh as a private gentleman and leader of Scotch natural philosophers. Among them he was equally at home as toastmaster, confidant, prankster and critic, as well as author of dissertations on rainfall or the progress of reason from simple folk knowledge to what then were the complexities of science.

Hutton had been born in Edinburgh on June 3, 1726. His father, a successful merchant and treasurer of the city, died while the boy was young. Mrs. Hutton thereupon decided that James should follow a profession rather than business, sent him to the endowed high school, and then to the city's university. Graduating while still a lad, he was "articled," or apprenticed, to a lawyer in preparation for a career at the bar.

But young Hutton had no liking for verbose writs, for pompous whereases and hereinafters, or for penalties by law provided to meet this or that offense. For one of his professors had told how simple acids dissolve ordinary metals, though gold will yield to nothing less than the mixture of nitric and hydrochloric acids known as aqua regia. The difference probably meant little to the professor, who used it to illustrate some principle of logic. But to Hutton it opened a glowing new world—the world of reactions and interactions between chemical substances. He read books on the subject while in college, and in back cubicles of the lawyer's office amused his fellow apprentices with simple experiments. Resulting odors must have reached more spacious front rooms, where the lawyer also heard exclamations not aroused by writs or prayers for judicial orders. After interrupting a series of such demonstrations, the disturbed employer "consented" to can-

Those Skeptical Scots

cellation of young Hutton's articles and advised the youthful scientist to seek some more congenial field.

Today such advice could be followed simply by change from law to chemistry. But one did not take up chemistry as a profession in 1744, just as no one became a physicist or a mechanical engineer. Medicine was the one recognized scientific profession, and as a medical student young Hutton returned to the University of Edinburgh. In 1747 he went on to Paris, where he worked at anatomy and chemistry with ardor for some two years more. Finding that Dutch universities were pre-eminent in medicine, he then left Paris for Leyden. There he passed the long examinations with honor, becoming a full-fledged M.D. in September 1749.

Degree and diploma raised the problem of a location for practice. Though not provincial, Hutton still was a Scot—a Scot who could not be happy when far from Edinburgh. Yet medicine in that city was controlled by a few eminent and old-fashioned doctors who united in opposition to new men of ability. Hutton saw that he would have to spend years of hard work for uncertain pay before he could build a practice that would put him among the leaders. Those years would use up his money, wear down his enthusiasm, take all the time and thought he might have given to science. What, he asked himself, would be their product? Another eminent but disappointed physician whose personal failure would make him bitter against doctors of a generation to come?

Hutton wrote his worries to James Davie, a chemist friend with whom he had experimented on production of sal ammoniac. Davie replied that the experiments were progressing and promised means for making the much-used ammonia salt out of ordinary soot. Perhaps a factory could be opened. Would it not be better to manage a chemical plant than to struggle against jealous doctors or set bones and deliver babies for people who could not pay?

Hutton thought well of Davie's plan, and then turned to face another problem. His father had left him a farm, smallish and none too profitable, but yet not to be sold. "What shall I do with such an estate?" asked Hutton. "Why not farm it yourself?" friends responded. "Try to farm with the same care and scientific knowledge you'd give to an operation or an experiment in chemistry. Nine chances to one you will make a good profit—and do something for Scotch agriculture, besides!"

A new and more complex experiment; the prospect appealed to Hutton. But farming demanded its own special skills, which in those days could not be studied at any university. Hutton learned that farms in Norfolk were superior, so there he set up as a tenant under hard-working John Dybold. He plowed, sowed, reaped and cared for livestock; having learned what Dybold could teach, he traveled through England, Holland, Brabant and Flanders in search of new methods. In Picardy he was delighted by the skillful cultivation of fields which were small, even by Scotch standards. Most English counties were not so advanced, nor were other parts of France.

In 1754 Hutton returned to Scotland, bringing a Norfolk plowman to do the heavy work in his fields. For fourteen years plowman and doctor worked together, using every recognized improvement and such new ones as Hutton could devise. Thanks to them the small estate became a show place; an example of experimental farming in a country where most field work was still governed by guesses and tradition. And for those who cared, Hutton's account books showed that this "huffing and puffing of science" brought a handsome profit. "More," said the owner, "than I'd ha' earned by struggling against yon fashionable doctors up in Edinburgh."

It seems probable that farming introduced Hutton to earth science. As early as 1753 he observed how surface soils were related to the underlying gravels, clays and hard rocks of various kinds. On his European trip he paid attention to minerals, which seemed to have close connection with the growth of crops. Later he toured North Scotland to study geology, discovered crystals in basalt, and noticed that rocks

Those Skeptical Scots

were dissolved by water seeping through them. He also noticed that rain water carried soil from fields and wore steep-sided gullies in bare banks and hills.

There was time for chemistry, too, both practical and "philosophic." The experiments with Davie progressed, and in 1765 the two men became partners in a sal ammoniac plant. It produced a biting white salt which then was a popular medicine in demand at a substantial profit. As farming methods caught up with knowledge of the day Hutton found work settling into routine which left still more time for science. In 1768 he decided to rent the estate to a well-trained tenant. He himself would move to Edinburgh and devote his life to what then was termed natural philosophy.

By this time Hutton was forty-two, a man of great simplicity but stubbornly established habits. He disliked social affairs and elaborate dress; except for his cocked hat with its small ornament he might have passed for a Quaker. Unmarried, he lived with three maiden sisters who ran his household like clockwork, leaving him free to write, make experiments, and spend long evenings with his cronies. Chief of these were John Playfair and Dr. Joseph Black, mathematician and chemist, to both of whom Hutton was deeply indebted for ideas. They often were joined by John Clark, an author on naval tactics. Another friend was Adam Ferguson, a historian whose works had been translated into both French and German.

Few records remain of the discussions in which these friends engaged. We know that they used "regularly to unbend themselves" in dinners of a society known only as the Oyster Club, where talk ranged from history to chemical problems, from popular prejudices to discontent in American colonies. At one meeting both Black and Hutton laughed at Scotch unwillingness to eat such things as snails. It was pure prejudice, they agreed, and to prove their point met together for a private feast of snail stew. Its odor was disturbing but they tried to eat, each thinking he alone was bothered. At last Black cleared his throat:

The Story of the Great Geologists

"Doctor," he ventured, "do you not think they taste a little —a very little—queer?"

"Damned queer!" roared Hutton, dropping his spoon. "Damned queer, indeed! Waiter, take them awa'!"

At another time several of these "highly respectable literary gentlemen" decided to meet for dinner once weekly, appointing Black and Hutton to choose the place. Entering a vintner's establishment, they asked to see the best dining room, engaged it, and there the meetings were held. All went well till one evening when Hutton came late, to be met by a bevy of young women who with meaningful glances and gigglings "took refuge in an adjoining apartment." Next day questions revealed that the learned company had been dining in one of the fanciest sporting houses of all Edinburgh!

Hutton was fond of walking, and trips about Edinburgh brought him face to face with geology. In the city itself was steep Castle Rock, a plug or boss of once-molten stuff that had hardened underground. To the east rose Arthur Seat and Salisbury Crags, both relics of eruptions in which lava had reached the surface. Westward ran the Water of Leith, foaming through ravines which showed the power of streams to wear away the land. In eroding, the Leith had cut long sections through tilted strata which Werner assigned to the Flötz, but which later would be assigned to the Carboniferous, or Coal Age.

The ex-farmer examined all these and then went afield for data. In Cheshire he studied salt mines; near Birmingham he visited iron mines and discovered beds of quartz pebbles which he traced into Wales. He also examined basalts and granites, considered beds of varied hardness, and developed opinions which would have outraged disciples of Abraham Werner.

For years, however, Hutton's work outraged no one because he did not publish. There might have been even more delay had he not taken part in founding the Royal Society of Edinburgh. At one of its early meetings he read a much-condensed statement of his opinions, a statement which had been

THEORY
OF THE
EARTH,
WITH
PROOFS AND ILLUSTRATIONS.

IN FOUR PARTS.

By *JAMES HUTTON*, M.D. & F.R.S.E.

VOL. I.

EDINBURGH:
PRINTED FOR MESSRS CADELL, JUNIOR, AND DAVIES,
LONDON; AND WILLIAM CREECH, EDINBURGH.

1795

TITLE PAGE OF HUTTON'S GREAT WORK

The Story of the Great Geologists

approved by both Playfair and Dr. Black. It was entitled *Theory of the Earth; or an Investigation into the Laws Observable in the Composition, Dissolution, and Restoration of Land upon the Globe*.

The essay was published in Volume I of the new society's *Transactions* and was attacked by Professor Richard Kirwan, of Dublin, who both accused Hutton of atheism and grossly misrepresented his work. The "impious" Scot had denied our earth a beginning, said Kirwan, and had made himself ridiculous by describing forces operating at the center of our planet. Equally absurd was Hutton's claim that coal was a mass of plant remains. Its origin, said Kirwan, was obvious; granite and basalt had decayed, producing grains of bitumen and carbon that settled upon the bottoms of seas. There they worked their way through soft clay banks and gathered in beds of coal.

Hutton read this attack while recovering from a long and dangerous illness. Next day he began to revise his reply, a full-length *Theory of the Earth* which friends had for years urged him to publish. It contained the original essay, followed by a more thorough statement of conclusions and the evidence on which they were based. It filled two volumes which appeared in 1795.

Hutton began by saying that the earth had changed often and greatly, a fact made evident by its rocks. Shale and limestone, for instance, lay under fields; valleys were cut through ridges of sandstone; conglomerates stood out in cliffs. But in spite of their elevated position all these rocks could be matched by deposits now gathering in the sea. To the aging Scot, conglomerate was nothing more than gravel or shingle cemented into stone, while sandstone was indurated deposits of sand. Most shale was mud compacted in layers, while limestone was largely made up of fragments worn from shells and corals. It added its bit of proof that much rock which now appears on land settled in the sea.

This conclusion might seem to support Werner; actually it

Those Skeptical Scots

opposed him. For Hutton's strata were no mere precipitates made of stuff that had swirled to and fro and then had settled in a primeval ocean. They were beds made of true sediment—of mud washed into bays by rivers, of sand and shingle once pounded by waves, of shells that were broken to bits by currents that rolled them along the sea bottom. Much of the finest material had been dissolved, just as water now dissolves limestones and minute particles from other rocks.

This difference led to others. Werner had described the oldest, or Primary, rocks as granites, schists, gneisses and slates which were the first precipitates to settle in his primeval ocean. To Hutton some of these were not sediments at all, and the rest were only hardened deposits of various ancient seas. Nor was he able to lump all rocks of one kind together as products of any special age. Puddingstone was made of gravel or shingle, and such material must have accumulated whenever and wherever waves beat against a rocky shore. Mud also must have settled in every age, as did grains of sand. Even beds of shells and corals, which seemed younger than most Primary rocks, still had piled upon sea bottoms at many different times. To crowd them all into one age or epoch would have been absurd.

Most sediments must have been loose when they settled, but many had long since turned into stone. Hutton explained this change by subterranean heat, which acted after strata were buried under later accumulations. In this he was largely wrong, for pressure, solution and chemical changes seem chiefly responsible. They, not melting, have turned mud into slate, have made marble out of limestone, and have produced the brown flints that are scattered through beds of English chalk.

Hutton also erred in supposing that heat had pushed deeply buried rocks upward into hills and mountain peaks. But he was right in saying that uplift had occurred, with breaking and tilting of some strata and intense crumpling of others. As a result some rocks stood on end, others were tipped over, and many were bent into folds or tight Vs. This doubtless had happened again and again, with convulsions that must have shaken great areas and ensuing periods of erosion. Thus, in Berwickshire, broken Primary strata stood on end beneath younger beds which were horizontal. Since these Secondary strata contained worn bits of the ancient stone, it was plain that frost, streams and waves had broken the uptilted strata into pebbles, sand and mud. The same thing, said Hutton, happens today as streams rush down mountains and waves beat on precipitous shores.

It is not clear that Hutton saw any connection between upheavals and thoroughgoing changes in the mineral make-up of rocks. But he did describe beds distorted in structure and crisscrossed by veins, with their "original or marine composition extremely obliterated." These rocks, he said, had been "more acted upon by subterranean heat" and had been "changed in a greater degree by the operations of the mineral kingdom." In our time such changes are recognized under the name of metamorphism.

On his trips about Scotland Hutton had seen other rocks which were not stratified and apparently never had been. Some were spread out over the surface; some formed masses like King Arthur's Seat; many were sheets called veins or dikes, which cut across other rocks of Primary and later age. Though some plainly were hardened flows of lava, others seemed to be once-molten stuff that had solidified under the ground. There they lay hidden until erosion stripped away their covering.

Hutton distinguished three groups of these once-molten rocks which had hardened after eruption or injection. Under the term "whinstone" he placed basalts, as well as other dark, heavy rocks now known as dolerite, diabase and andesite. Porphyry included a variety of lighter rocks with large, angular crystals in a finely granular groundmass. His granite included the coarse-grained rock known by that name today, as well as a few other kinds which now have names of their own.

Whinstones, said Hutton, were closely related to lavas that now flow from volcanoes. Those that lay in sheets were

Those Skeptical Scots

erupted, but dikes and irregular masses or plugs had been forced into older rocks which might later be worn away.

Granite was lighter and coarser than whinstone; instead of lying in sheets or dikes it formed masses many miles in length and thousands of feet in thickness. It also seemed to underlie or cut across very much older rocks. Yet the fact that it did cut across them proved that it was younger than they and had been forced into weak zones or cracks. Had not Hutton found a series of such granite veins along the river Tilt, rejoicing over them so loudly that guides felt sure his discovery was gold?

Hutton died in 1797, before his third and last volume was ready for the printer. Yet his "theory" lost little by that, for never did a man state great ideas in more cumbrous and less inspiring style. Thus his idea of metamorphism—thrilling when reannounced forty years later—was packed into one sentence of 136 words arranged in almost unmanageable clauses. Hutton's great discoveries might have come to nothing had not his stanch friend, Playfair, put them into a small and readable book.

John Playfair was born in 1747, son of a Forfarshire preacher. In boyhood he showed genius for mathematics, but after study in Aberdeen and Edinburgh became a minister in the town of his birth. In 1785 he went back to Edinburgh as professor of mathematics and there was led into geology by walks and talks with Hutton. He devoted his holidays to geologic trips in Great Britain, Ireland and on the Continent, visiting volcanoes of France and Italy and glaciers of Switzerland. He probably was the first man to realize that huge "erratic" boulders, far from their parent granites, might have been transported by glaciers which since had melted away.

Playfair's book, called *Illustrations of the Huttonian Theory of the Earth,* caused a sensation when it appeared in 1802. In terse, often dramatic, sentences it presented Hutton's evidence and drove toward his conclusions. "If the coast is bold and rocky," wrote Playfair, "it speaks a language easy to be interpreted. Its broken and abrupt contour, the deep gulphs

The Story of the Great Geologists

and salient promontories . . . combined with the inequality of hardness in the rocks, prove, that the present line of the shore has been determined by the action of the sea.

"Again, where the sea-coast is flat, we have abundant evidence of the degradation of the land in the beaches of sand and small gravel; the sand banks and shoals that are constantly changing; the alluvial land at the mouths of the rivers; the bars that seem to oppose their discharge into the sea, and the shallowness of the sea itself." If we proceed inland from the shores, Playfair added, we find that elaborately branched valleys of streams "have been cut by the waters themselves . . . and that it is by the repeated touches of the same instrument, that this curious assemblage of lines has been engraved so deeply on the surface of the globe."

Having traveled widely and observed for himself, Playfair was able to add to some of Hutton's discussions and give prominence to several subjects which the latter had not emphasized. Specially able were accounts of uplift and folding of strata, the intrusive origin of granites, and erosion along the shores of lakes. Playfair also had studied raised beaches along the Scottish seacoast and used them as evidence of uplift rather than retreat of the sea by actual lowering of its waters. Thus he gave the first convincing account of oscillations in the level of European land.

Such treatment made plain what Hutton had obscured: the fact that his "theory" was no mere set of speculations about the origin of our planet and changes within its interior. It was, instead, a system of earth science based upon existing conditions, and on processes that operate today. He first asked how modern sediments originate, how lands are being worn away, and how they are given shape by the agents of degradation. Not until such questions were answered did he try to reconstruct the past.

Earth's present features and changes explain its past. Hutton based his work upon that thesis, and with Playfair's help made it both a fact and a guiding principle of earth science.

CHAPTER VI

Neptune versus Vulcan

HUTTON HAD PUSHED earth science a long way forward when he explained erosion and sediments, proved that rocks had been uplifted, and showed why the earth's present was a reliable key to its past. But Werner's followers were ready for none of these advances, nor for Hutton's work on vulcanism. What, they asked, was this nonsense about heat deep within the earth, with molten rocks that hardened underground and others that spread widely over the planet's surface?

Not nonsense but science, answered Playfair, and thus conflict began. Before long it divided geologists into two camps whose members battered each other with words as often as they broke stones with hammers. Vulcanists, who followed Hutton, insisted upon the importance of heat in the hardening of sediments as well as in surface eruptions and rocks like granite, which formed underground. Werner's pupils, known as Neptunists, scoffed. Had not the great inspector declared that the earth's center was cold, that basalt and granite were precipitates, and that volcanoes were trifles warmed by burning coal and peculiar to recent times?

The argument began in books, but it soon was taken directly from Freiberg to the classrooms and "whinstone" cliffs of

The Story of the Great Geologists

Scotland. For after two years of study under Werner, Robert Jameson returned to Edinburgh in 1804 as professor of natural history in Hutton's alma mater.

Jameson was a remarkable man who combined stiffness and a tightly closed mind with personal charm, vast earnestness, and rare ability as a teacher. Within a short time he gathered a band of ardent students in courses modeled on those of Freiberg. In 1808 he organized a Wernerian Natural History Society with the master himself as an honorary member. The Society's *Memoirs* were open to papers which followed every Freibergian rule. Lest those rules longer remain in doubt, Jameson wrote the *Elements of Geognosy* as part of a larger work on what still posed as mineralogy. The *Elements* forthwith became the one authoritative printed source of Neptunist doctrine.

Werner taught and sat stodgily at home; Jameson traveled Scotland from its southern counties to the distant Shetland Isles. On each trip he saw things with Freibergian eyes, unconsciously distorting facts to fit his preconceptions. Coming home, he wrote out his discoveries, taking care to sneer at "fire philosophers" and praising Werner, whose teachings bore no resemblance to "those monstrosities known under the name of 'Theories of the Earth.'" With confident faith he discussed excavations which were convex, told how basalt sheets had settled from water, and explained that pumice could not have come from volcanoes, since it was stratified. Unable to dispose of the water in Werner's primeval oceans, his faith in them remained unshaken. "Although we cannot give any very satisfactory answer to this question," he maintained, "it is evident that the theory of diminution of water remains equally probable. We may be fully convinced of its truth, and are so, although we may not be able to explain it. To know from observation that a great phenomenon took place, is a very different thing from ascertaining how it happened."

No wonder an English critic complained that Wernerian doctrine obstructed progress.

Neptune versus Vulcan

By supposing the order already fixed and determined when it is really not, further inquiry is prevented, and propositions are taken for granted on the strength of a theoretical principle, that require to be ascertained by actual observation. It has happened to the Wernerian system, as it has to many other improvements; they were at first inventions of great utility; but being carried beyond the point to which truth and matter of fact could bear them out, they have become obstructions to all further advancement, and have ended with retarding the progress which they began with accelerating. . . . It is not so much to describe the strata as they are, and to compare them with rocks of the same character in other countries, as to decide whether they belong to this or that series of depositions, supposed once to have taken place over the whole earth; whether, for example, they be of the Independent Coal or the Newest Flötz-trap formation, or such like. Thus it is to ascertain their place in an ideal world, or in that list of successive formations which have nothing but the most hypothetical existence:—it is to this object, unfortunately for true science, that the business of mineralogical observation has of late been reduced.

Conflicting schools have a way of settling upon some special problem, determined either to solve it or fall. The Vulcanists might have made another choice, but *Wernerismus* took the origin of basalt, that dark, hard and often columnar rock which Guettard had called a precipitate and Desmarest had proved to be lava. Neptunists sided with the former, Vulcanists with the latter, and for decades the battle raged. It brought the Freibergians their first great defeat—and that at the hands of two men who began as faithful geognosts.

Jean François d'Aubuisson had spent four years as a favored student of Werner, enjoying long hours together while the inspector explained details of the system which he could not bring himself to write. The pupil then prepared a book on Saxon basalts which maintained that their origin as precipitates was not to be disputed. Any contrary opinion was due to man's "love for the marvellous. . . . It delights the imagination to suppose the existence, in former ages, of burning mountains and great streams of liquid fire, nature being, as it were, in conflagration at the very spot perhaps now occupied by our peaceful homes. The black and sooty appearance of basaltic masses, and the isolated and spectacular shapes of mountains which they form, greatly help this illusion."

When these conclusions were presented to the Institute of Sciences in Paris, they brought a suggestion that the author examine districts where volcanic records were more convincing than they seemed to be in Saxony. "The Citizen d'Aubuisson knows how to observe," wrote two critics, but he "has seen neither active nor extinct volcanoes. Since he has lived until now in the midst of water-laid formations, we should like him to visit regions where fire has displayed its power."

The author was impressed but not discouraged; witness the fact that he published his book before taking a coach for Clermont and the famous *puys*. The trip was pleasant but promised to be a wild-goose chase, for the very first basalts D'Aubuisson saw rested upon a thick mass of granite. This, by Werner's rule, was a Primitive rock; one that had settled a long time before the Flötz with its coal accumulated upon the earth's surface. Since there was no coal to burn and melt overlying rock, the basalt could not have begun as lava.

That seemed to settle the question—but did it? Scarcely had D'Aubuisson proved the general presence of granite when he discovered dozens of bare cones with craters and "cinders" of pumice on their slopes. He traced flow after flow of lava which had breached the rims of craters and had rolled down into valleys or merged with basaltic sheets.

The facts [he found], spoke too plainly to be mistaken. . . . I must either have absolutely refused the testimony of my senses in not seeing the truth, or that of my conscience in not straightway making it known. There can be no question that basalts of volcanic origin occur in Auvergne and the Vivarais. There are found in Saxony, and in basaltic districts generally, masses of rock with an exactly similar groundmass, which enclose exactly and exclusively the same crystals, and which have exactly the same structure in the field. There is not merely an analogy, but a complete similarity; and we cannot escape the conclusion that there has also been a complete equivalence in formation and origin.

Neptune versus Vulcan

D'Aubuisson courageously read his retraction before the French Institute of Sciences in 1804, publishing soon afterward. But so persistent was Neptunism, and so uncertain were the means of scientific communication, that an Edinburgh edition of the original *Basalts of Saxony* appeared in 1814. The translator did remark that M. d'Aubuisson was said to have modified his opinions on the basaltic deposits of Auvergne, but apparently had said nothing new about those of Saxony. The latter, therefore, might still be accepted as columnar precipitates.

D'Aubuisson saw his error, corrected it, and thereafter wrote books which removed him from the ranks of faithful geognosts. Other workers, more conservative than he, followed Werner as long as they could and then made such halting, piecemeal corrections that they seemed to distrust their own minds. This course, indeed, was taken by the man whom some critics still regard as the greatest geologist produced by Germany.

Christian Leopold von Buch—he dropped the first name—was the sixth son of an Austrian diplomat and minor nobleman. Born in 1774, he took a short course in chemistry and mineralogy at Berlin when he was only sixteen and then attended Freiberg. There he lived almost three years in Werner's home, receiving even more intensive instruction than had been given D'Aubuisson. Von Buch then went to the universities of Halle and Freiburg and served a year as an inspector of mines in Silesia. The work was useful but detailed and far from inspiring. Having inherited a "competence for life"—it was hardly a fortune—he resigned with the plan of spending his time in independent research.

This work began in 1797, on orthodox Wernerian lines. Von Buch toured the Alps with a friend of Freiberg school days, walking from inn to inn and drawing sections intended to prove that strata tilted in ridges and snowy peaks had settled as they stood. Next year Von Buch went alone through Austria to Rome, examining near-by lavas which evidently had been molten. In 1799 he reached Vesuvius, which seemed far from trifling as he climbed to it over remnants of a still greater volcano long since blown to pieces. Other volcanic fields were found to be both large and ancient. In them the young geognost saw basalt columns in obvious lava flows and discovered undoubted lavas whose crystals should have made them precipitates in the Primitive sea. But Freibergian doctrine somehow survived; when Von Buch published a book on his travels it was dedicated to Werner and stanchly defended the aqueous origin of basalt. "Every country and every district in which basalt is found," it declared, "gives evidence against any idea that this remarkable rock was erupted in molten condition."

Thus he wrote early in 1802, but before the words got into print Von Buch visited the Auvergne. Like D'Aubuisson, he climbed the black-rimmed plateau, traced columnar sheets till they merged with ordinary lavas, and followed those lava streams to the craters from which they had flowed. Yet the German was not one to change his mind quickly or to make sweeping recantations. Convinced that the Auvergne basalts had been molten, he said so in letters, but would not extend that conclusion to rocks near Naples and in Germany. For seven years he delayed, considered arguments pro and con, weighed this or that objection—and then confessed himself bewildered. German basalts seemed to be volcanic, but *"opinions* stood in opposition which only new evidence could remove."

Von Buch's account, published in 1809, was an amazing mixture of clear thinking, apparent ignorance, and evasion. He dwelt upon the perfection of each French *puy*—"a veritable model of the form and degradation of a volcano, such as cannot be found so clearly in either Etna or Vesuvius." Then, almost unbelievably, he declared that these models were to be found in a region "about which the naturalists of France have talked so much . . . but which they have never yet described," as if neither Desmarest nor Guettard had put his

Neptune versus Vulcan

work into print. Some of Desmarest's conclusions were even credited to a lesser author whose *Essay on the Volcanoes of Auvergne* was said to be "an excellent work."

Now came the carefully rationalized evasion. As if making a confidant of the reader, Von Buch declared that conclusions which were sound in the Auvergne could not reach into Germany. Auvergne's volcanoes and lava flows lay on granite; in Saxon mountains things were more complex. Geognosts must "remember how many different kinds of rocks are there associated with basalt as essential accompaniments and how they form, with basalt, a connected whole which absolutely contradicts any notion of volcanic action."

We can almost see the writer's mental recoil from an alarming conclusion. *Since basalts and stratified rocks intermingle, this is not a matter of one single kind of rock. If, therefore, I yield on a single point do I not undermine Werner's whole doctrine of chemical precipitations? But that is unthinkable! There is—there must be—a difference between basalts of France and Germany!*

From France Von Buch traveled to Norway, whose formations then were all but unknown. He crossed western Lapland with its stormy lakes, its valleys leading down to deep fiords, and its knobs of gray and pink granite scraped bare by glacial ice. But ancient glaciers were unsuspected in those days, and Von Buch's eyes were on the rocks. Slowly he followed them southward, taking boat from island to island and from inlet to inlet, but going ashore to study each formation and ticket it in Werner's system. All went well till he neared Christiania (now Oslo) where tongues and oblique sheets of granite reached upward through beds of fossil-bearing limestone which were crumpled and baked near the contact. Von Buch faced the same facts Hutton had discovered: veins of granite that must have been molten as they forced their way through hard though fractured rocks. And instead of being Primitive in age, the granite was obviously younger than Flötz limestones which it had changed.

The Story of the Great Geologists

Again Von Buch was challenged, and again he made reservations which avoided an open break with Werner. A more serious rift, in fact, appeared when Von Buch decided that much of the Swedish coast was rising from the sea, for this was the kind of change Hutton had used to explain the uplift of normal marine sediments. Some years later Von Buch announced that mountains in central Europe had not settled under salt water but were built by a series of rifts and upheavals that turned former sea bottoms into high land. Each upheaval caused torrents that scoured valleys, washed blocks and fragments of stone downstream, and swept boulders from one mountain to another. Sometimes it even tossed them through the air across newly scoured depressions.

Those boulders were the very ones which a young Swiss named Agassiz would attribute to glacial action. But the man who had treasured Wernerism had no patience with such nonsense. In 1847, ten years after Agassiz offered proofs of a vanished ice age, the great German marched from boulder to boulder, striking the biggest ones with his cane and demanding, "Where is the glacier that could have moved this great block and left it sticking here?" He also ridiculed Darwin's coral islands "dancing up and down in the sea" and denied that lava flows ever had built volcanoes. Those mountains, he said, arose like bubbles which sometimes burst and sometimes remained intact. For such origin he had "complete proof," and let no mere upstart doubt it!

For Von Buch was an authoritarian as well as a conservative. He cherished familiar ideas; he believed that the great men of science did work which no ordinary mortal might question without proving himself an ignoramus and perhaps a fool. Novel theories were heresy, and heresy in his eyes was a form of intellectual sickness. *"Glacial* Drift, the Glacial Epoch!" he once fumed. "It was once a frightful disease in Switzerland, a kind of wide-spreading cholera; it passed lightly over Germany, but went over and has fixed itself on the other side of the Channel."

Neptune versus Vulcan

Yet a vast curiosity drove Von Buch onward—to the British Isles, back and forth across Europe, through Russia and to the Canary Islands with their volcanic peaks. In each region he was compelled to limit some dogma or make some advance which kept him near the front of his science, though not among its pioneers. Not till old age overtook him did it become plain that he took each step forward with hesitation and was ready to turn accepted innovations into revered antiques.

By that time, however, no one cared, for Von Buch had become a beloved and eccentric giant from the scientific past. Younger men smiled at his buckled shoes, his knee breeches, his long black silken hose. But they were delighted when he came hundreds of miles to see them, with no luggage except a green cloth bag into which were stuffed a fresh pair of stockings and one clean white shirt. Every host knew that Von Buch would stay with him at least till the shirt newly put into use could be laundered and take its place in that bag.

Even more amazing were his field trips, also in silk hose and low shoes, with only an overcoat as protection from snow or rain. At seventy-three Von Buch trudged Alpine paths in a downpour, refusing offers of a carriage. On reaching shelter he called at once for dinner, being much too hungry to waste time changing his clothes. He ate heartily while water oozed from his shoes, snorting, *"Nein! Nein!"* when someone offered dry woolen stockings in their place. On another trip he walked twelve hours through alternating drizzles and downpours, yet when evening came he joked and told stories. Next morning he was up at five o'clock, ready for another day of grinding mountain travel.

One could let such a man treasure the past, even when he lamented Neptunian days which his own hesitating advances had helped bring to an end.

CHAPTER VII

Like Goes with Like

Werner had divided earth history into three great ages and one of minor importance, each represented by a series of rocks which overlay or overlapped one another in order as they had formed. Hutton recognized a comparable though much less orderly sequence, since beds had been changed, broken, uplifted and partly worn away. Neither he nor Werner had any reliable rule for determining the age of individual strata, nor could Huttonians arrange scattered formations in one continuous series. Yet the means for doing so had been worked out; not by a philosopher or rule-ridden geognost, but by a country surveyor and engineer who was self-taught in science.

William Smith was born in 1769, eldest son of an "ingenuous mechanic" who died before his son was eight years old. When Widow Smith married again the boy was sent to live with his father's brother.

Uncle William was a crotchety farmer, unmarried and with small respect for anything except manual labor. He was willing that his nephew attend village school, since anyone should read, write and do sums or simple multiplication. But he objected vigorously when the boy collected "pundibs"—fossil

Like Goes with Like

mollusk shells—and the petrified sea urchins locally known as poundstones. To the man both of these were nonsense unless, of course, one was a dairymaid and could use the poundstones to weight covers on earthen crocks of cheese.

Nonsense, too, were books on geometry and surveying which Nephew Will wanted to buy. But the boy begged and wheedled until he got them, plunging earnestly into their maze of theorems and procedures. The elder William was somewhat mollified when he found that theorems helped to measure fields, and that figures and sightings which looked foolish could be used in draining bogs. Yet he would not consent to praise or encourage, but stubbornly dragged his nephew from study to good, productive work on the farm.

At eighteen young Smith became a surveyor's assistant, moving to his employer's house. For three years he held the rod and dragged the chain—but he also observed different kinds of soil, noted "agricultural and commercial appropriations" of the land, and studied borings for coal in the New Forest as well as in other regions. At twenty-two he began independent work which led to a detailed "subterraneous survey" of some important coal mines. It showed that certain rocks always lay in a definite order, a fact which the tradition-bound miners stubbornly refused to accept. Smith kept faith in his own conclusions and awaited a chance to use them.

That chance came because Brindley, an engineer, learned to lay out canals as waterways for heavy traffic. They caught public imagination, for England was astir in spite of wars, heavy taxes, and the threat of greater wars to come. Rich men and poor alike were aware that steam had been put to work in coal mines, bringing out quantities of fuel for new factories and smelters. Would these not mean more plants, more business, more jobs? Would not the canals help out by allowing coal to be hauled far and cheaply, as it could not be hauled in wagons?

Smith saw a future for canals and also for canal builders. He saved money to buy more instruments, sat up late to study new texts, and at twenty-five became both surveyor and engineer for the still unauthorized Somerset Coal Canal. While laying out its course he examined beds, or strata, which without exception were tilted southeastward and lay one above another like "superposed slices of bread and butter." Yet the position of these stony sandwiches did not match that of the buried coal strata below, which were tilted at various angles between great fractures now known as faults. Though Smith could not realize it then, this discordance was the result of an earth revolution during which mountains were raised, deltas were shaken, and lowlands were turned into seas. Though not catastrophic, it was much more real than the sudden upheaval with which Von Buch would try to build the Alps.

Parliament approved the Somerset Canal in 1794. With two members of its governing committee—we should call them directors—Smith promptly set out on a nine-hundred-mile journey to study the construction, management and traffic of other inland waterways. They saw one with a tunnel four miles long, visited silk mills near Birmingham, and stopped at mines where long blocks of coal were piled up for sale like cordwood. They admired steam hoists at other mines and noticed horse-powered railways whose tracks were made of wood or wood plated with iron. They also remarked that a surprising number of structures were poorly designed—"inconvenient and unphilosophical," to use the young engineer's words.

The trip was made in a light carriage, or chaise, whose folding top was put down except on rainy days. Smith rode in front with the driver, watching formation after formation reach the surface as the carriage rolled northward. He found, too, that the surface gave a clue to rocks beneath it; that rolling meadows meant chalk while damp pastures with willow trees were underlain by clay. Stops to examine bridges, locks and retaining walls enabled him to check details that could only be glimpsed from the chaise.

For six years Smith was engineer of the canal, surveying each segment of its course and supervising construction. The

Like Goes with Like

work was crammed with detail, yet it encouraged just those geologic observations which he wanted to make. He had to notice the lay and the character of rocks, seek clues by which to recognize them, and predict formations to be encountered as the waterway grew longer. In all these tasks he paid close attention to fossils, which most people then dismissed as mere curios. Smith found them far more distinct than thick beds or thin; than colors dark, light and mottled; than grains of sand or fine sediment composing this or that stratum. More important, he found that, although different formations might look alike, each had its own peculiar remains. Though Smith still did not know what these were, he soon learned to recognize their shapes. Before long he was identifying one formation by its thick poundstones, another by pundibs or petrified nuts, and so on through the series of rocks along the Somerset Canal.

This achievement posed a new problem. If fossils distinguished different formations, should not rocks be grouped together when their petrified remains were the same? Suppose gray shale in one hill contained two odd kinds of pundib, and that shells of the same sort were found in limestone a good many miles away. Should not both rocks go into one formation, even though they were unlike?

To answer this question Smith tramped over hundreds of hills, followed uncounted banks and examined dozens of canal cuts. He traced fossils and beds from one to another, checking both variations and similarities. He proved that remains did permit the matching of beds, and he also learned to correlate series of beds which might lie miles apart. Suppose, for example, that shale near the crest of one hill contained horn-shaped shells identical with those near the base of a distant quarry. Smith made the two shell zones one and so secured a sequence, or section, that ran from the hill's foot to the top of the quarry.

So far, so good; but what came next? Smith watched workmen dig in a bank beside the canal, where a limy clay revealed

The Story of the Great Geologists

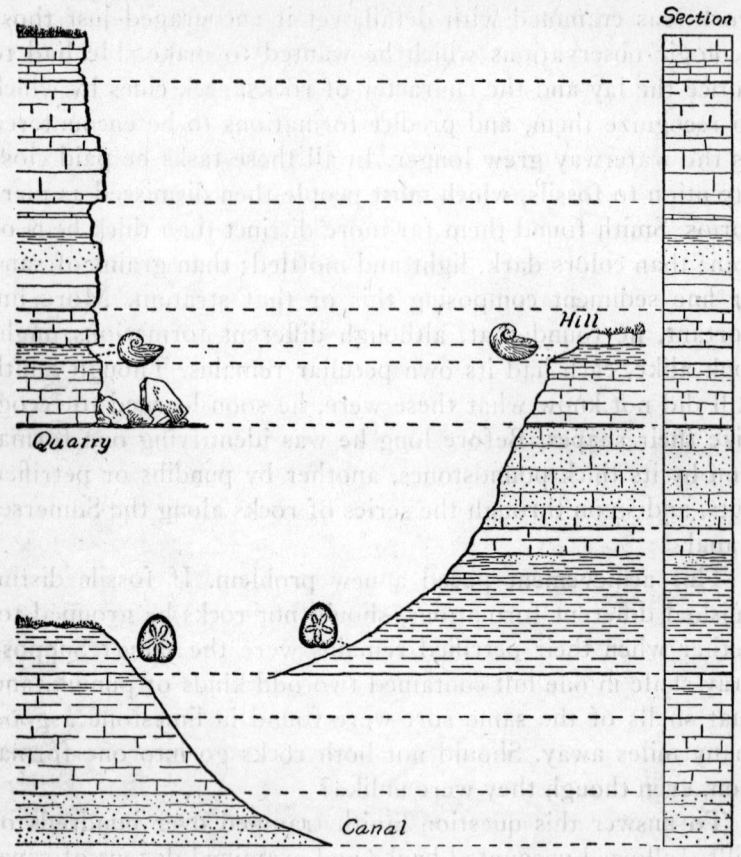

Diagram showing how William Smith used fossils to match beds in a quarry, a hill and a canal cut. When combined they give the section at the right.

pundibs like some low down in the hill. Again beds with similar fossils were matched so that strata exposed below the pundibs were added to the bottom of the series. In diagram the process can be represented by three sections that overlap. A fourth section shows the beds of all three in continuous sequence from the oldest, which lies at the bottom, to the youngest at the top.

Here was a great discovery; one that promised order for

Like Goes with Like

the tangled geologic record. The promise inspired Smith, who by that time had married and moved to a cottage in the city of Bath. There, during leisure hours of 1796, he planned a book that would explain his method and give the general sequence of British rocks lying above the Coal. Readers who followed his rules could fit any formation into this sequence. Since it persisted from place to place, they also could tell what kind of rocks would lie above or below any identified formation. Specifically, a man who found the pundib bed in a half-covered slope would know that below it lay rocks seen in the canal cut while above it were lower beds of the hill on which oysters had been found.

Smith planned the book well but put it aside when the Coal Canal demanded more attention. He also was discouraged when no one to whom he talked thought much of his discoveries. Many smiled, a few argued, some changed the subject. One steward of an estate did warm up a little. Smith's methods were interesting, he said, and they might—only *might*—be useful. Would they tell what some unfamiliar field was like and what it might be good for? Of course, one still must plow and put in crops to make sure of the soil's value. . . .

Thus things went for three dreary years, till Smith called upon a minister who collected fossils. Made near Bath, the collection still was a hotchpotch, for the Rev. Benjamin Richardson knew little or nothing of strata. He was amazed when Smith picked up specimen after specimen, told where it had been discovered, and in just what bed or formation. Still more amazing, the stocky visitor announced that "the same strata were found always in the same order and contained the same peculiar fossils." Such a claim was not merely radical; it was quite unknown to savants throughout the geologic world!

Perhaps, rumbled Smith, but it was true. The way to prove it was to collect more fossils and see whether they followed rules he would set down in advance. The Rev. Richardson agreed, and to strengthen the test called in Joseph Townsend, a fellow divine and author. Smith pointed out a distant hill; it

The Story of the Great Geologists

should be capped with freestone containing "petrified rams' horns," or ammonites, which Richardson had collected in similar beds near Bath. The fossils were found and the trio went on, testing stratum after stratum till the skeptics were convinced.

It was one thing to check Smith's conclusion, but another to persuade him to put it into usable form. The subject de-

The Rev. Benjamin Richardson,
from a silhouette.

served a volume, but he had little time for writing and no money for publication. At last a dinner led to a compromise: a brief "tabular view of the subject" which could be handed about to geologists in lieu of a publication. Smith dictated while Richardson wrote, supplying technical names of fossils to replace vernacular terms. The table listed formations from Coal to Chalk, noted the presence of springs, and contained such comments as "A rich manure" for liming fields, or "Visible at a distance, by the slips on . . . hills around Bath." Within two years this "Card of the English Strata" was copied and freely given to geologists of Germany and other countries of Europe.

Like Goes with Like

The Rev. Joseph Townsend,
after a drawing by William Smith.

By this time the surveyor-engineer-stratigrapher had grown into a "stiff and stubborn fellow" of thirty, cordial to all who agreed with him but determined to follow his own ways despite hardship or opposition. He had quit the Somerset Canal in a huff; unwilling to be a jobholder again, he set up as a consulting engineer. In this role he drained marshes, built sea walls, controlled landslides and laid out irrigation ditches as

The Story of the Great Geologists

well as waterworks. When Bath's best springs failed he restored their flow; when a landlord complained of worthless fields Smith proved that under them lay valuable coal beds. He drained some canals, "puddled" others, and located sources of water to fill necessary locks. As his fame grew he raised his prices from one guinea per day and "extraordinary" expenses in 1799 to two guineas and all expenses in 1801 and three guineas—about fifteen dollars—several years later.

Those were high wages at a time when good meadowland could be had for £3, or $14.50, per acre. Yet Smith did not become wealthy nor even well-to-do. He bought a small estate near Bath, but sold a bit of inherited land and spent every available shilling to collect fossils, examine series of strata, and record them on his maps. If all went well he labored till the last minute and then bounced off by special coach to some engineering appointment. Drivers envied his power—and chance—to sleep while the vehicle lurched through the dark over rutted country roads.

Meanwhile the book begun in 1796 was postponed again and again. A prospectus appeared in 1801; eminent men subscribed to the work and a duke promised his patronage. But the nobleman died, the publisher-to-be failed, and Smith went on with engineering to pay for further field work. In 1805 he rented a house in London for the display of his fossils and hired an artist to draw them for reproduction, only to give up when he found that publication would cost three thousand pounds. That expense, he wrote a supporter on June 26, "is too great for my circumstances . . . and there seems to be a tardiness among many of the great personages who were expected to subscribe very liberally." The solution, he thought, was to publish in parts, and for these a new prospectus would "shortly be printed."

This plan went the way of others, and as year dragged after year Smith's friends began to worry. He had scattered his "Table of Strata" far and wide; originals had been copied as they passed from hand to hand. Contents therefore were pub-

Like Goes with Like

lic property; anyone who wished might revise them a bit and print the revised version as his own. Indeed, copies then in existence probably had passed through so many hands, with so many changes, that even their owners might not know from whom they originally had come.

More important, earth science was marching on and other men were approaching Smith's conclusions through their own researches. Take that French ex-botanist Lamarck, for example; in a book about fossil shells he had said that they furnished clues to ancient earth revolutions. Two other French scientists had examined hills and quarries near Paris, where strata could be distinguished by petrified bones and teeth of extinct mammals. At last, in 1808, the newly formed Geological Society of London concluded that Smith had abandoned his plan for a stratigraphic map of Britain and asked its president to prepare one. He was getting data wherever he could find them and was progressing with speed.

This last fact may have spurred Smith to action; perhaps he was cheered by a publisher who wanted to bring out a geologic map and asked no subsidy. At any rate, Smith began work in January of 1813, only to be halted by the problem of naming different formations. His uncertainty delayed the engraver, and so did new trips to examine coal mines and survey drainage operations. A year went by, and most of another, before the sheets showing southern England were printed and laboriously colored by hand. They were exhibited before the Board of Agriculture late in 1814, while sovereigns of the Allied Nations were visiting London to celebrate Napoleon's defeat. Board members may have wanted to watch foreign royalty, but they loyally listened to Smith as he talked of formations and outcrops. In the end, one member of Parliament was so stirred that he subscribed a generous sum for himself as well as fifty pounds in the name of his deceased father-in-law. The president of the Royal Society and other wealthy men also "contributed to soften the dire aspect of utter poverty" which Smith's life by that time presented.

The Story of the Great Geologists

The complete map appeared in 1815 under the staggering title of *A Delineation of the Strata of England and Wales, with part of Scotland; exhibiting the Collieries and Mines, the Marshes and Fen Lands originally overflowed by the Sea, and the Varieties of Soil, according to the Variations in the Substrata; illustrated by the most descriptive names.* Prices ranged from £5 5s. to £12 ($25.50 to $58.30)—enormous sums by modern standards, but not too much for a well-engraved map measuring six feet by eight feet six inches, on which twenty colors had been applied by hand. The lowest part of each formation was deeply shaded, a device which helped to demark each one from its subjacent neighbor and also kept novices from seeing formations as if they were upside down.

Some four hundred and fifty copies of the map were sold, in four issues whose coloring varied to fit new facts discovered by Smith himself or provided by other workers. Neither publisher nor author profited greatly, and Smith felt that he must secure an income without daily work. He therefore built a railroad across his land near Bath, to open quarries and haul building stones to the Somerset Canal.

Perhaps Smith was hasty or prejudiced; perhaps he had bad luck. At any rate, the stone's quality failed, and with it the costly railroad. Smith had to sell his land, his furniture, his books; he gave up the house in London and even sold his best fossils for a lump sum of six hundred pounds. They went to the then new British Museum, which advanced another hundred pounds for arranging and cataloging. Smith himself became a wanderer who for seven years had no home except inns and furnished rooms.

Yet he neither gave up geology nor complained; indeed, he worked with more determination than he had shown in 1803. During 1816 he produced four sections of a seven-part work on *Strata Identified by Organized Fossils*, with plates printed on paper whose shades matched those of the rocks in different formations. Next year came the *Stratigraphical System of Organized Fossils*, a work based upon the collection sold to the

Like Goes with Like

British Museum. Six "geological sections and views" came in 1819, and in 1820 a small *New Geological Map* based on the large one, but with better coloring. Between 1819 and 1824 Smith also published six parts of a *New Geological Atlas of England and Wales,* which was "calculated to elucidate the Agriculture of each County, and to show the Situation of the best Materials for Building, Making of Roads, the Constructing of Canals, and pointing out those Places where Coal and other valuable Materials are likely to be found."

While these publications made their appearance, their author labored on as an engineer, took geologic trips on foot, and relaxed in villages so remote that he failed to get an urgent summons from Russia, which needed mineral "inspectors." In 1821 he began to lecture for scientific associations; four years later, officers of the Literary and Philosophical Society of Kingston-upon-Hull reported that Smith had delivered a series of nine lectures for fifty pounds. Admission tickets repaid all but twelve pounds and nine shillings, and since that sum was more than offset by new memberships, the lectures really were "enjoyed by the Society gratis."

At Scarborough Smith went collecting with local fossil hunters, found a new water supply for the city, and became friendly with a baronet who laid plans for improving the estate which he would soon inherit. In 1828 he became the baronet's land steward, and for six years served as a combination of agent, manager and agricultural engineer.

The work was light and left much time for writing; beyond doubt Smith's employer planned it so in the hope that he would revise his yellowing bundles of notes and turn them into publications. But the habits of decades were too strong; the engineer who had rushed from one job to another could not settle down to the task of turning notes into articles or volumes. Instead he filled more bundles of paper with sketches and "memoirs," and then blamed their disorder upon the demands of stewardship. In 1834 Smith resigned, saying that he had grown "weary of nothing but farming concerns" and needed his last years to complete and publish manuscripts. The baronet offered no objection, but provided a retaining fee of twenty pounds per year for occasional advice and examination of fields.

With old age came scientific honors. Smith once had been forced to defend his claims, printing a booklet which told of the exhibit in London, the freely distributed "Table," and the use of his method by geologists. Now he was accepted as a pioneer—as the father of stratigraphy and inventor of geologic maps. In 1831 the Geological Society of London gave him its gold medal; next year came a pension from the Crown, whose ministers had been prodded by eminent scientists. The pension was a hundred pounds; enough to guarantee comfort for a man of simple tastes.

Smith also was made much of by the young British Association for the Advancement of Science. Bristol made him its guest at the meeting of 1836, paying all his expenses and giving him a chance to see his old friend, Richardson. Smith also spent several days in examining a coal field, putting his discoveries on another hand-colored map. Two years later he toured England and Wales with an official geologist in search of stone to be used in the Houses of Parliament. Smith continued this work into 1839, rested and walked near Scarborough and took a boat for London in July. He then went overland toward Birmingham, where the British Association would meet late in the month of August. Having time to spare, he traveled slowly, visiting friends, collecting fossils and following formations whose age was still in doubt. On one side trip he "caught cold," though descriptions of the attack suggest that it really was influenza. He went to bed, weakened rapidly, and died on August 28.

It would be easy to grow sentimental over this man; to call him a neglected genius and imagine what he might have done had he held a professorship or possessed an independent income. But Smith himself anticipated such laments with realistic dissent. Friends, he wrote, had been many and generous—

Like Goes with Like

so generous that they worked to help him, provided money, and even published his books while he was away from London. But the labor that took so much of his time? It was just what he needed, since it forced him to study strata and fossils, putting his knowledge of them to use. As for leisure—well, Smith knew his own defects in spite of a few attempts to deny them. He liked to muse, delay, digress; nothing but the need for money would make him stick to a job. Books brought no profit, so he put them off, but the need to live kept him hard at the field studies from which would come his fame.

Travel, books, work with other geologists—these, too, would have been of doubtful value. Other men had journeyed through Europe; Smith's discoveries came from close study of strata within a limited region. Books would not have helped in this, except to name the fossils for which he used farmer's terms. Nor would geologic companions, though their friendship would have lessened the sense of isolation which was a burden for years.

Finally, it is well to remember that Smith made his great discoveries before he was twenty-seven. While such leaders as Von Buch still clung to methods of the past, he was proving that strata were orderly and was using that order to build a system of stratigraphy and to prove the value of fossils as marks of geologic age. At thirty he made a geologic map which both traced formations at the surface and showed what became of them when they sank underground. Though his later maps were more detailed they contained no essential concepts not found in the first one.

In short, Smith was one of those lucky men who do their great work when young. Then with stubborn persistence he elaborated by filling in series, coloring maps, checking fossils in stratum after stratum. Such work strengthened and applied his theses until they commanded respect from every geologist. Since Smith's work made that strength possible, we cannot look upon it as a grievous disadvantage or waste.

CHAPTER VIII

Knight of the "Principles"

Forfarshire, once called Angus, lies between mountains and the eastern coast of Scotland, where firths trace the courses of rivers now sunk beneath the sea. The shire is crossed by a great earth break, or fault, whose course determines a valley in which Pict tribesmen once came from their underground houses to battle invading Romans. Later wars may be traced by Danish spears or in roofless walls and battered towers that survive from medieval keeps. Castles in time gave way to manors, whose gabled additions still tell of growth through succeeding generations.

One rambling manor, called Kinnordy, was owned by Charles Lyell. Unlike Smith, he belonged to the landed gentry —a laird whose income as well as home was derived by inheritance. With no need to earn, he took landowning lightly; to study plants or read verse pleased him more than the task of supervising farms. Lyell collected mosses and lichens, won repute by describing them, and then made a second reputation by translating poems of Dante and less famous contemporaries.

But both Dante and lichens were far out of mind on the fourteenth of November 1797. On that day Charles Lyell,

Knight of the "Principles"

"gentleman of means and refinement," fidgeted like any ordinary mortal while a son—his first child—was born in one of Kinnordy's rooms. Neighbors remarked that the baby should have a fine disposition, since he came at the onset of a winter which was unnaturally warm. So mild did the weather stay, indeed, that Mrs. Lyell dared to sleep all night with her windows open!

Of the infant Charles we know only this: that teeth came through his gums very slowly, leading one neighborhood crone to predict that he would grow up toothless. He also ignored the omens of a sunny temper, crying so much that he "was pronounced to be the loudest and most indefatigable squaller of all the brats of Angus." Yet he must have done much of his wailing beyond that county, for his parents soon went to southern England in search of a pleasant climate. There they moved from place to place until Lyell senior leased an acceptable house some six miles from Southampton. During these wanderings another son, Tom, was born as second in the full assemblage of three sons and seven daughters.

Charles learned the alphabet at three, an achievement which became his first clear memory. But almost five years passed before he and Tom were sent to a boarding school at Ringwood, some eighteen miles from Southampton. There they and three other lads were the youngest in a group of fifty rough-and-tumble youngsters.

Ringwood was a coastal town in a district where smuggling still was a profession. The boys played contrabandist versus customs men until chances for greater excitement came with the threat of a French invasion. Even the elder Lyell gave up lichens and Dante long enough to serve as captain of Defense Volunteers with a company stationed at Ringwood. The Volunteers saw no service except a test alarm, but they were on hand to celebrate when the French fleet was defeated at Trafalgar. The Lyell boys ran from bonfire to bonfire and watched Volunteers throw firecrackers through the windows of a mayor who had refused to let his fields be used for a drill ground.

The Story of the Great Geologists

The school at Ringwood was not very good, nor did its students have high social standing. So Charles and Tom were transferred to Salisbury, where the master was famous for his learning and pupils came from the best families of three adjacent counties. Charles, however, resented their hazing, disliked the bare, walled-in playground, and was bored by walks in which the boys marched along suburban sidewalks. He was glad when illness took him home, where he spent much time outdoors collecting butterflies. A servant gave him sympathy and help, but the family laughed at his "bug hunting" as an unmanly pursuit.

Poor health made schooling intermittent; the boy who knew his alphabet at three had passed his nineteenth birthday before he was ready to enter Oxford. There he studied classics and mathematics—the usual program—but also did some real work on insects with one of the college masters. Having learned a little geology from a textbook in his father's library, he also attended a course given by William Buckland.

Buckland was one of Oxford's most popular professors; a cheerful, bustling and eloquent teacher whose special concern was hyenas, elephants, and other ancient beasts whose bones were buried in British caves. He wrote a large and popular book on the subject and constantly carried a blue bag from which, "even at fashionable evening parties, he would bring out and describe with infinite drollery . . . the latest 'find' from a bone cave." At a meeting devoted to fossil footprints he acted out the part of a chicken that stalked about making tracks in ordinary farmyard mud. He delighted in jokes about his work, and sent friends engraved copies of a poem which professed to announce his death. His corpse, said the author, would be encrusted in cave lime and then set up as Buckland's one suitable monument.

Buckland described the argument over basalt with such fire that Lyell used his first long vacation to examine volcanic rocks in Scotland. With two friends he climbed the lava cliffs near Edinburgh, crossed the high Grampians, and took the boat for

Knight of the "Principles"

Staffa to see its then newly discovered columns. The trio rowed into Fingal's Cave, admiring the roof made of broken pillars while waves washed their boat to and fro. This seemed less of a hazard than an inn that was nothing more than a hut with a smoky fire built on the floor and walls decorated by dried fish tied in evil-smelling pairs.

Buckland encouraged Lyell's geologic interests; they were fixed by a three months' tour of Europe in the summer of 1818. While his parents and sisters took easier trips, Charles made the ordinary climbs, walked across the Mer de Glace, and traced the course of Alpine floods. The guide who helped him over the glacier paused for lunch at a meadow where violets and gentians bloomed a few yards from moving ice. After eating, Lyell slept on the ground, contrasting this bed with accommodations at inns where fleas and lumpy mattresses deprived tourists of rest.

Lyell spoke his mind about the fleas, but did not let them take his attention from things that daily changed the land. On one mountainside water soaked into clay; the result was an avalanche that brought countless tons of rock down upon fields and a village. A lake formed behind a dam of ice; when the dam weakened it loosed a flood that covered roads and houses with rubbish, swept hillsides bare of soil, and rolled massive boulders downstream. Even melting snow became violent as its water foamed down cascades and plunged over falls. In many places it whirled stones round and round, grinding them as they scoured deep potholes in ledges of massive stone.

Lyell took his degree at Oxford in 1819 and to please his father studied law with a "special pleader" in London. Failing sight twice sent him out of the office and to France, where he listened to lectures on science, met eminent men, and declined invitations to the opera because the bright lights hurt his eyes. The year 1825 at last saw him accredited to the bar, following a circuit court and so spending half his time at law. During the remainder he studied rocks and fossils, attended meetings of the Geological Society of London, and wrote a popular article which defended Playfair and Hutton. He finally gave up legal work in 1827.

By that time Lyell was a pleasant, squarely built man of thirty whose brow seemed too wide and high for his face. To his eyes all distant objects were blurred; he stooped to see people clearly and sometimes could not tell where to step. Despite this handicap he was planning a book which would amplify the ideas of Hutton and organize the diverse essentials of geology. Such a book demanded new data as well as a careful review of regions already described. Lyell therefore planned a long trip in continental Europe with his friends, the geologic Murchisons.

Roderick Murchison was another well-to-do Scot, five years older than Lyell. He had entered the army when only fifteen, had campaigned against the French in Spain, and was commissioned a captain at twenty. Six years later he married; his wife was a general's daughter who urged him to resign when Napoleon's defeat killed hopes of military advancement. She planned a cultured life with attention to literature and art, but Murchison took up fox hunting with both gusto and endless attention to details. He seemed settled in a life of trifling when Sir Humphry Davy urged him to try the mental niceties and adventures of science.

Mrs. Murchison seconded Sir Humphry, for she was tired of horses, dogs and red coats. Murchison dallied more than a year, hunting, visiting and playing the social lion among manors in southern Scotland. But in 1824 his wife won out; the ex-soldier sold all except two horses and began to attend lectures on geology and chemistry. He heard Lyell read his first technical paper and at the age of thirty-three published one of his own. Within three years he followed it with extensive field work, planned and began a series of important memoirs, and (as evidence of respect among his fellows) became a secretary of the Geological Society of London.

Such was the man whom Lyell now chose as a summer companion. Mrs. Murchison wanted to go along, not so much for

Knight of the "Principles"

the trip as to share her husband's work and keep watch over his digestion. The party left Paris in May 1828, with both men on the box of an open carriage while Mrs. Murchison and her maid rode behind. Lyell's secretary—his former law clerk—had been sent ahead to look for plants, collect insects, and do other scientific chores.

The rig broke down while still in the city, almost at the door of the carriage maker who had been paid to put it in first-rate condition. More collapses were threatened on the rutted highways worn out by generations of traffic. Beyond Moulins the roads were not so bad; the horses could make better speed and wheels could roll instead of bouncing into and out of holes. The geologists began to trace ancient lake deposits, follow hardened lava streams, and examine extinct volcanoes. They found one which Desmarest had missed, dug up fossils for Mrs. Murchison to pack, and encountered mountaineers who maintained that Napoleon was still in Europe. They worked from five or six in the morning until sunset, with Murchison urging early starts, bargaining at inns, and eating enormous meals which resulted in constant indigestion. These he fought with a diet of pills while cursing the steady heat. Damn a land where the mercury climbed above eighty every day in July!

If the Auvergne appeared hot, southern France was hotter, and the air of Italy seemed ablaze. It sent both Murchisons off to the Tyrol, but Lyell stayed to climb volcanoes and to cross fields of cinders, study beds of sand mixed with sulphur, and trace the results of ancient earthquakes. On the island of Ischia he collected sea shells two thousand feet above salt water; proof that the island had risen that much in relatively recent times. Elsewhere he examined marine deposits sandwiched between clays that had settled on the flood plains of streams. He also made notes on political persecution; upon foodless markets, half-starved children, filth and discouraged ignorance. No reformer, he tried to take things as they came, but rebelled when a servant stored meat in his shoes and a landlord offered a bedchamber which was nothing more than a lean-to against the more substantial mule stall. "Each of the dirty family sleeps in it [the chamber]," he wrote, "and turns out for the stranger!"

Lyell returned to England before March 1829, and quickly got to work on his book. But there were collections to study, problems to discuss, conclusions to be tested with care before they were put into writing. Despite the author's high hopes and his publisher's prodding, Volume I of the *Principles of Geology* was not ready for printing till the spring of 1830. The second volume came out in 1832, and the third in May 1833. By that time the work was in such demand that Volumes I and II had been revised and a new edition must be prepared for the entire set. "My book has got on so rapidly," wrote Lyell, "as to put me in great good humor with the public."

Not all his public, however, was in great good humor with Lyell. Though Werner was dead and Moro forgotten, men still treasured fallacies and resented anyone who opposed them. Thus Buckland kept faith for a while in Noah's Deluge, adding two other world-wide floods that carved valleys, sent boulders floating like corks, and spread sheets of "diluvial" gravel over lowlands and hills. Von Buch and the French geologist, De Beaumont, continued to write of upheavals by which whole mountain ranges were raised almost overnight. Other authors dwelt on catastrophic eruptions, as well as earthquakes followed by torrents that rivaled Buckland's floods. Every outburst remodeled the earth, swept it bare of living things, and prepared it for others to be furnished by new divine creations. The result was a sequence of periods divided by cataclysms and distinguished by plants and animals that appeared when each age began and vanished as it came to an end.

To such scientists Lyell seemed a wild speculator, a willful iconoclast who followed Playfair and Hutton. Earth's present, he insisted, was the key to its past; to decipher events of ancient times we must learn what is happening now. And once the effects of rains, waves, earthquakes and volcanoes are understood they account for past revolutions, leaving neither

Knight of the "Principles"

need nor room for world-shaking catastrophes. *No causes whatever,* Lyell maintained; no causes whatever have changed the earth except those that still do so under the eyes of man. Nor have these causes run down in modern times, for they still show as much power and speed as they displayed during ancient ages.

Such blunt statements were made in letters to friends; in his book Lyell took a course of conciliatory caution. He first traced the growth of geology itself, showing how knowledge of the earth had grown despite errors that popped up again and again. He examined the "prepossessions" from which those errors sprang, removing the basis for false though cherished speculations without attacking them in detail. From this he moved to problems of climate, and readers who began with curious history suddenly found themselves aware that neither science nor faith had need for a past whose conditions were fundamentally unlike those of modern time.

All this was preparation; now came the real argument. It began with modern earth processes and their work, a survey so crammed with facts that it still serves for reference after more than a century. No one who read it could deny that rivers, waves and volcanoes change the land or that forces which often result in earthquakes carry it up and down. Nor could many deny that these alterations might be made with speed. Did not the Chilean shores rise four feet in a day, turning oyster beds into land? Had not waves cut the coast of Yorkshire away at a rate of seven to fifteen feet per year since the Norman conquest? And what of the harbor of Sheringham, where water twenty feet deep covered the spot on which a fifty-foot cliff had stood in 1781?

Still, it was one thing to show that the earth was changing, but another to prove that such changes had not reached cataclysmic force. To do this Lyell reviewed the Tertiary Period, or Age, whose fossils had been used to prove a series of worldwide catastrophes followed by new creations. Actually, said Lyell, the Tertiary could be divided into four parts, or epochs, distinguished by the percentage of shells that *survived through succeeding times and still were living in modern seas.* The last "upheaval," which had pushed much of Europe above salt water, took place with so little disturbance that most of Lyell's Sicilian fossils belonged to species still found in the Mediterranean. Yet Sicily itself had risen as much as three thousand feet!

To complete his argument Lyell traced modern processes in rocks dating from ancient times. He showed that some strata plainly had formed on land, just as dust and sand now settle in beds or in belts of dunes. He recorded lava flows and eruptions of ash or pumice, noted breaks that must have resulted in earthquakes, traced beaches now lifted into land, and described the deposits of long-vanished hot springs. He noted a soil bed below the Portland stone and traced bays, bars and swamps in formations of the Coal Age. He even found traces of currents and waves in strata whose antiquity is now estimated at eight to ten million years. Though younger than most geologic records, they strongly supported Lyell's belief that earth processes have persisted from ancient to modern times.

As we have seen, the *Principles* demanded a new edition even before it was finished, but Lyell was no man to sit in a study slaving at one book. While Volume I still was in press he dashed off to the Pyrenees, where he found rocks filled with marine shells and examined ripple marks much older than those he had seen in England. In 1831 he was appointed professor of geology at King's College, in London; he also made another trip through continental Europe and wrote travel journals for Miss Mary Horner, to whom he was engaged. They were married at Bonn a year later and geologized on their honeymoon.

Lyell's lectures at King's College began in the spring of 1832, for a class in which ordinary students were outnumbered by scientists, lawyers and men of affairs, with their sisters and wives. Each talk was prepared with meticulous care, and the speaker himself paid an artist to paint appropriate diagrams.

Knight of the "Principles"

Murchison was enthusiastic; the course was "talked of over London"; it set social belles to studying science for the first time in their lives. But this last result did not please the college, which rewarded Lyell's success by closing his lectures to women. When his second-year class dropped to fifteen the disappointed professor resigned.

In 1841 the Lyells sailed for their first visit to Canada and the United States. Landing at Halifax, they went to New York, took a boat up the Hudson to Albany, and traveled overland to Niagara Falls. Lyell took great delight in the country, praising its scenic geology, its citizens who treated newcomers as old friends, its bustling innkeepers who gladly got dinner for six or eight people in only ten minutes. He also liked the pioneering geologists who showed him their fossils, gave him notes on their unpublished work, and took him to places that should be mentioned when the *Principles* would next be revised. In such company he examined thick series of strata, visited coal fields, and traced successive faunas in the greensands of New Jersey. In the Appalachians he made the surprising error of concluding that the mountains were folded "when in a soft and yielding state." Neither he nor his new friends realized that hard rocks bend readily when under great compression and weight.

Geologizing was interrupted while Lyell gave a series of lectures for the Lowell Institute in Boston. This engagement had brought him across the ocean; it was planned for twelve non-technical talks to which tickets were issued by lot. But when 4500 of them were taken it was agreed that the "class" should be divided, with repetition of each lecture. Accustomed to social distinctions and a sometimes rowdy proletariat, the Englishman was delighted by an audience in which rich men and poor mechanics mingled, as did their daughters and wives. He remarked that members of every group conducted themselves with the "utmost decorum"—and listened to what he said.

Lyell also lectured in Philadelphia, after spending the winter between Richmond, Virginia, and Charleston, South Carolina. He later studied dinosaur footprints in red sandstones of Massachusetts, traveled westward to Cincinnati and Cleveland, and revisited Niagara Falls. Crossing to Canada, he collected modern species of sea shells on hills in Montreal and noticed that old marine beds in Mount Royal were cut by basaltic veins. In Nova Scotia he watched tides rise fifty feet, their waves battering rocks from cliffs that once were swamps of the Coal Age. Stony tree trunks now stand in the bedded headlands, with their roots in gray shale that began as the water-soaked soil of bogs.

Back in Britain, the Lyells went to Kinnordy and then hurried southward to London. They had been away thirteen months; there were notes to arrange, papers to write, and three dozen big boxes of fossils to be labeled and put into collections. Not till these urgent tasks were done could Lyell take time to write his *Travels in North America,* a book consisting of two small volumes which were published in 1845.

The *Travels* shows Lyell as a man concerned with human affairs as well as with the data of science. In England he had nurtured reforms; in America he found those reforms growing out of New World conditions. Wages were high because men were scarce; houses were improved because people had to build new ones; education was liberalized because everyone had a share in it and the clergy did not exercise control. So large did this factor loom that Lyell devoted much of his book to a plea for sweeping reforms in the universities of England.

He was not, of course, blind to defects: to the menace of slavery in the South, to the North's abolitionism, and to political cheapening that came with universal suffrage. But the balance was favorable, and even a rent rebellion and financial crisis seemed far less alarming to him than to critics who had viewed them from Europe. The rebellion came from one overlarge estate poorly administered and in conflict with general trends. The crisis was no worse than financial bubbles that had burst in old and seemingly stable England. Lyell took pains to

SEDGWICK
eighty-two, when the
nbrian controversy was
ning.

After Clark and Hughes

SIR RODERICK MURCHISON
in middle age, the leader
of British science.

LOUIS AGASSIZ at fifty-five — Harvard's most famous professor.

American Journal of Science

After "N

WILLIAM MACLURE from an engraving companying Morton's graphical sketch of

Knight of the "Principles"

show that most states were solvent and had built great public improvements with their own—not borrowed—money. If some, such as Pennsylvania, had to suspend payments on debts, it merely meant that many people still were poor and that poor people voted for acts which brought them immediate relief. If England's poor also could go to the polls would they not demand lowered taxes? And would they care much if that reduction meant default on foreign-owned bonds?

By 1846 Lyell's collections had become so large that their owner had to leave his apartment and take a spacious house. There his wife could entertain while he spread out specimens and worked on a new edition of what had become "the book."

For the *Principles* had become a career; a book that lived, changed form, and grew with the progress of earth science. Starting out as a treatise on earth's changes, it added facts about the past and in time became a general work that recorded research, digested it, and shaped it into a unified whole. The fifth British edition had grown to four volumes, but the last one soon became an independent work on earth's history and rock formations. Now the three remaining volumes were combined into one, with large pages and many illustrations. Lyell dedicated this edition to his father-in-law, then president of the Geological Society. A copy was presented at the anniversary meeting of 1847, with speeches in which colleagues praised Lyell and he praised the United States. A clergyman lamented that his fellow divines, quarreling about trifles, lacked the poise of scientists who could dine in friendship even when they disagreed on world-shaping theories.

One of these disagreements arose when Darwin offered his theory of evolution. Lyell had been cool toward earlier work in this field, which leaned too heavily on dogmatic speculation. But Darwin was a cautious man—so cautious that friends had to insist that he publish the *Origin of Species* without excessive delay. Lyell took the book with him on a trip to study ancient footprints, wrote an enthusiastic letter about it, and announced himself quite ready to extend Darwin's conclusions to man.

The Story of the Great Geologists

Within a few months he was at work on another edition of the *Principles,* in which evolution would receive fifteen thorough and favorable chapters.

But this new edition was not needed at once, and Lyell's mind was captured by the notion that man had evolved and so had roots reaching back into ancient times. At the age of sixty-three, therefore, he put geology proper aside for three years in order to study relics of early human races, as well as bones of the animals with which they were associated. He tramped along terraces of the Somme, went through caves of the Meuse, and measured sections in English gravel pits where workmen dug up fossil elephant teeth and axes of weathered flint. Skilled in summarizing the work of others, he was able to publish the *Antiquity of Man* in February of 1863. The book captured public attention, and entered its third edition before the end of the year.

Study and writing made such demands that Lyell refused to accept time-wasting offices. Yet he always was glad to investigate mine accidents, rescue neglected works of art, or take part in international exhibitions for the good of industry and trade. He also became a leader in the movement to reform British universities, a movement that achieved success some years after his *Travels in North America* attacked conservatism and waste.

On public committees he repeatedly worked with Prince Albert, husband of Queen Victoria. Yet this association seems to have had no connection with his knighthood, an honor that came in 1848. Lyell was deeply pleased by it but not flattered, and combined the formal visit to Balmoral Castle with "geological exploring on the banks of the Dee, into which Prince Albert entered with much spirit."

Lyell's judgment of royalty, indeed, was warped neither by hero worship nor British symbolism. In his eyes Victoria was an able queen, an understanding wife, and a good mother. Her husband stood out as a man of wide knowledge and many interests who combined high position with an alert social con-

Knight of the "Principles"

science. One could work with such a man and admire him; when the Prince died in 1861 Lyell's grief was personal. Two years later he visited Victoria at Osborne, where he found courtiers reading the *Antiquity of Man*. The Queen already had gone through her copy, and instead of showing conservative alarm, was ready with questions about caves, ancient races, and Darwin's theories. "She has a clear understanding," Lyell wrote his wife, "and thinks quite fearlessly for herself."

Lyell's eyesight grew little worse with the years, but he suffered attacks of illness that marked the onset of old age. Still, neither he nor Mrs. Lyell could settle down to inactivity. They entertained friends, revised books, traveled; after Lyell's first illness they toured Scotland and the English coast, going on to the Alps. The summer of 1872 found them at Aurignac, in a region rich with remains of Stone Age man. They planned for Switzerland again in 1873, but Mrs. Lyell suddenly fell ill and died on April 24 of that year.

That quick death was a shock to her husband, yet he did not collapse. "I endeavor," he wrote, "by daily work at my favorite science to forget as far as possible the dreadful change in my existence." Three months later he was busy with a new edition of a textbook as well as with preparations for the postponed trip to Switzerland. In 1874 he made a last trip to Scotland; on November 5 he took part in the half-century meeting of a dinner club whose members also belonged to the Geological Society. As one of the club's founders he spoke with a vigor that delighted his friends.

With December Lyell's health declined; he could not leave his room on New Year's of 1875. Death came on February 22, and friends quickly petitioned for his burial in Westminster Abbey. It was, they said, Britain's only proper resting place for the "most philosophical and influential geologist that ever lived, and one of the best of men."

CHAPTER IX

The Cambrian Conflict

THE DALE OF DENT begins as a gorge below a mountain pass in Yorkshire. In lower country the gorge becomes a valley wide enough to hold stone-walled farms and a crumbling village which oldsters still know as Dent's Town. It once was an isolated but prosperous place, famous for hand-knitted woolen stockings and firkins of strongly salted butter. Its one paved street ran between stone houses whose galleries covered outdoor stairways and in summer shaded women who sat at whirring spinning wheels.

March 22, 1785, was too cold for outdoor gossip and spinning, but not for the midwife and surgeon who hurried to the house of the Rev. Richard Sedgwick. While the surgeon cared for Mrs. Sedgwick his helper tucked a newborn baby in her apron and scurried to a back parlor where "the Reverend" was trying to write. "Give you joy, sir!" she bubbled, spreading the apron. "A fine boy he is, sir, and as like you as one pea is to another!" Then she gasped as the worried father retorted, "Like *me*, do you say, Margaret? Why, he's as black as a toad!"

The swarthy youngster, named Adam, grew up like the sons of less learned dalesmen. He tore his books in mischief and

The Cambrian Conflict

then was sorry; he attended the village school, played leaping games among gravestones, and begged rides on his father's mare. At fourteen he was given a gun, poached a bit without being caught, and went fishing in the public stream. He also gathered nuts, not getting so many as other boys because he sometimes wandered off to collect fossils from beds of limestone in the hills.

Though young Adam had learned to read at five, he was almost twenty years old when he finally entered Cambridge. By that time Napoleon's wars had reduced the freshman class to 128 and seemed to have turned all England into a military camp. The young man marveled at guns that lumbered past his coach, at columns of foot soldiers, and at cavalry on prancing horses. He also noticed that civilian trade was almost paralyzed, that prices were exorbitant, and that coaches which bumped and swayed left even a young body exhausted after three days and two nights. He was happy to spend the third night in a stationary bed.

Prosperous students smiled at Sedgwick's poorly cut clothes and noticed his Yorkshire accent. But they applauded in June of 1805, when a stiff examination placed him near the head of his class, and were sympathetic in the autumn, when he fell ill of typhoid. The attack was so serious that he could not leave his room before the following spring.

Sedgwick's plan had been to get a general education and then "read" law in an office. But his father's health began to fail; unable to count on money from home, he first secured a scholarship and then became a university fellow with the ministry as his goal. But the strain of study was too much for a body still weakened by that long bout with typhoid. For five years Sedgwick took long rests in the country, tutored small groups of students, and studied theology when he felt unusually well. For a while Napoleon's triumphs so depressed him that he resolved to emigrate to the United States if the French should conquer England. Instead, he was the first man in Dent to learn of Waterloo, reading the news aloud to assembled villagers. A month later he returned to Cambridge as assistant tutor in mathematics.

By 1816 Sedgwick felt he could afford a summer in continental Europe. Paris was spoiled by his prejudice against its Catholicism; indeed, he soon felt a "cordial hatred of all the ways and works of the French." He did enjoy Holland, Germany and especially Switzerland, where he rode along mountain paths and was amazed by the scenery. He was almost as much surprised to find that monks could be kindly, hospitable and polite; the prior at St. Bernard was as pleasant and well informed a man as travelers could wish to meet. "We rose next morning at four," Sedgwick wrote back to England, "and were astonished and not a little pleased to find two honest monks up, with some warm coffee and toast, to see us off." That word "honest," from a Dalesman's pen, meant that the monks were good, wholesome men.

Sedgwick cared little for mathematics and no more for theology. But Cambridge then was controlled by the Church, with rules which compelled tutoring fellows to join the clergy or lose their positions. This threat drove Sedgwick on with his studies; he was ordained in July 1817, and preached a few sermons at Dent during his summer vacation. Within a year, however, he found tutoring so distasteful that he applied for the endowed professorship of geology. Elected by an overwhelming vote, he wondered just how he would live on a wage of one hundred pounds, with ten pounds more as an allowance for field work, experiments, correspondence and purchase of specimens.

No one could have known much less of geology than the newly elected professor. But those were times when a classical education was supposed to provide the foundation for all sorts of intellectual work; special knowledge or skill might be picked up quickly to meet this or that special need. Sedgwick read a little and then went to the country, where he examined strata, collected fossils, and pried into the secrets of lead mines. As professor he would go down a thousand feet, crawl through

The Cambrian Conflict

dripping tunnels, load his knapsack with heavy specimens. As a still youngish man released from care he then would come to the surface, change his clothes and attend boisterous miningtown balls as master of ceremonies.

From the lead mines Sedgwick went to copper shafts of Staffordshire and the great salt mines at Northwich. Back in Cambridge he prepared a successful series of lectures and then, though he might have rested, rushed off on another trip. "That lively gentleman Mr. Sedgwick," as one stranger called him, had stumbled into a calling which aroused his enthusiasm.

This second trip was cut short by an accident, but in August Sedgwick was out again. He rambled along the coast of Somerset and tried to see not rhyme but reason in structures of the Mendip Hills. "They afford," he wrote, "fine specimens of the contorsions exhibited by that rock to which geologists have given the name of greywacke." With irony he added: "What a delightfully sounding word!"

He might have exclaimed, "What a jumble to put under any name!" For the Greywacke, or Grauwacke, was an assemblage of sandstones, slates, quartzites and schists which Werner had put together when he named the Transition Group. He supposed them to be products of that intermediate age during which animals started out in the sea and worn sediments were beginning to replace primeval precipitates. As Jameson put it, they recorded the "passage or transition of the earth from its chaotic to its habitable state."

Sedgwick shared this opinion in 1819, for his self-taught earth science began on strictly Wernerian lines. But within the next five years he discovered lava sheets among sediments and began to realize that the Greywacke was merely a theoretical catchall for rocks which no one understood. Some were stratified and some were not; some had fossils but others were barren; many were bent into great flexures or were sliced and shattered by faults. Though they seemed unrelated to other formations there were hints that they must be assigned to several series which differed greatly in age.

The Story of the Great Geologists

Sedgwick reached these conclusions in England's Lake Country, where he made a detailed geologic map and became a close friend of Wordsworth. Later he joined forces with Murchison, the dashing and wealthy ex-soldier who was

Sedgwick on a field trip, from an ink sketch.

Lyell's partner in France. Working in unison though separately, they determined to find both order and stratigraphic limits within the "interminable Greywacke."

Their methods offered a contrast—but not the one we expect. Sedgwick, the cautious fellow who had turned preacher to save an unattractive job, at once made for the highest, most rugged part of North Wales. There he rushed about in battered stovepipe hat and flapping coat, looking like Ichabod Crane as he traced strata, matched faulted blocks and devised three great new formations which had no known relationship to rocks of established age. In his haste he even ignored fossils, basing his formations upon characters of beds and their mineral grains.

The Cambrian Conflict

Murchison, still the handsome ex-officer and playboy, proceeded with meticulous caution. No tramping alone over wind-beaten wilds; no rushing into complex regions far from geologic base. After much planning he set out for South Wales with his "wife and maid, two good gray nags and a little carriage, saddles being strapped on behind for occasional equestrian use." On the way he visited geologists, noting down every fact they could give him about the Greywacke. Then came rambles with William Smith's nephew, on which the whole problem was put aside while the pair smoked Murchison's costly cigars and discussed the relationships of limestones. Weeks passed before the carriage rolled into southern Wales, where Mrs. Murchison made sketches while her husband did more active work.

There were two sound ways to do such a job: to begin at the bottom or start from the top and carefully work through the series. Murchison had the good luck to find Greywacke just beneath a well-known "Secondary" formation still famous as the Old Red Sandstone. Not only did the beds lie in order; the uppermost beds abounded in fossils whose stratigraphic limits were known. "For a first survey," Murchison reported, "I had got the upper grauwacke, so called, into my hands."

Through three more summers he worked, adding bed after bed to the series reaching downward from the Old Red contact. At last he felt that the strata needed a name, partly to distinguish them, and partly to show that they were linked in one seemingly continuous system. Since they were exposed in country of the ancient Silures, Murchison called them Silurian. He also urged Sedgwick to name his three formations, which appeared to form another system lower in the geologic column. Why not call them the Cambrian, from the Roman name for Wales?

The friends described their new systems jointly in 1835. Next year they attacked another problem, that of "greywackes" in Devon and Cornwall which might be Carboniferous or might antedate the Cambrian. They were incredulous when fossils hinted that neither possibility was true. Introductory studies were followed by two years during which sections were measured, strata were mapped, and fossils were studied by skilled paleontologists whose verdict became more and more convincing. By 1839 Sedgwick and Murchison were ready to tell the Geological Society of London that another system, the Devonian, lay between the Silurian and the Carboniferous. Its base was the same Old Red Sandstone seen in southern Wales.

This new change outraged conservatives, who organized to repel an epidemic of stratigraphic division. Sedgwick was ill and busy at home, but rumors of the coming attack permitted his collaborator to prepare both evidence and rebuttal. With the skill of a lawyer he led critics on and then got admissions, till the opposition found itself supporting what it meant to defeat. "It was right well that I was *not* absent," wrote Murchison, "or things might have gone *pro tempore* against us." The conservatives were routed instead, and friends of the new system "looked upon the case as settled" by that one well-managed debate.

The Devonian thus went its way in peace, but problems awaited to plague the Silurian and Cambrian. From the first both systems had been imperfect: groups based upon rocks exposed in different regions, without known overlap and with nothing but inference to tell which was the older. Although the Silurian had been firmly grounded on fossils, no one knew how deeply those fossils might range before stopping at some stratigraphic break. Further difficulty lay in the fact that Sedgwick's three Cambrian divisions were based only on mineral grains and general appearance of strata. Such characters were notoriously inconstant; how much would they mean should fossils be discovered?

Several years went by without difficulty; years during which the friends described their Devonian System, Sedgwick grew more and more unwell, and Murchison plodded ahead with detailed stratigraphic studies. Their first result was *The Silurian*

The Cambrian Conflict

System, a fat volume of eight hundred large pages plus an atlas of plates and sections, as well as a colored map. It divided the Silurian into upper and lower series, described their formations in detail, and dealt with the principal animals that "had peopled the waters in which those early deposits were laid down." The book was dedicated to Sedgwick, who found it a magnificent present which, however, he refused to review. "I have not time," he explained in a friendly note, "to write to newspaper editors any letters fit to be read."

Murchison continued to extend his Silurian, with results followed by scientists of the new Geological Survey. By 1840 he began to find "Cambrian" fossils of species identical with those known from his own Lower Silurian. Two years later he asked permission to say that the two systems were one, since "Professor Sedgwick has now satisfied himself that the lowest organic remains which can be traced are . . . Lower Silurian."

By accepted rules of correlation, this meant that fossil-bearing parts of the Cambrian would be known as Silurian. Sedgwick, however, did not think of that; he gave his consent and then was amazed when Murchison made the transfer. For years he nursed a growing wrath, becoming more and more angry as other men followed Murchison's example.

At last Sedgwick's resentment burst out—not in a staid and acceptable correction, but in words so harsh that the Geological Society tried to suppress them after they were in print. Next year, in a popular guide to the Lake Country, Sedgwick charged that Murchison had misrepresented his claims, had given "strange" explanations, and that he harbored a notion of the Cambrian "not derived from anything I had ever said or written." A new book, *Siluria,* brought further charges of misrepresentation and a demand for apology that could not be satisfied. In 1861, when Cambridge gave Murchison an honorary degree, his one-time friend and collaborator was absent from the ceremony.

It is hard to understand such a quarrel, which Sedgwick

[105]

sometimes denied, sometimes explained away by ill health, and sometimes carried on with vindictive bitterness. Its real basis, perhaps, was an obscure complex in which lack of money, sickness and disappointment fused with jealousy and even envy.

Unlike Murchison, Sedgwick was poor; his professorship paid a starvation wage which had to be supplemented by preaching. From 1834 onward he spent part of each year as prebendary at Norwich, living in a queer old house and conducting services in a gloomy church. He also suffered from varying ailments—gout, indigestion, nervousness and vague symptoms which even his friends sometimes traced to mental rather than physical ills. Reinforcing his native reluctance to write, these troubles kept his discoveries hidden for years in notebooks or boxes full of specimens. Then, when Murchison published articles or books, Sedgwick felt neglected if not victimized.

The hurt was deep, yet accusations that followed were unjust and imaginary. Sedgwick's letters show that he approved Murchison's plans in advance, that he held opinions which were later disavowed, and that he joined in decisions which were blamed upon Murchison alone. Such tactics might have brought reproof, but long after Sedgwick's charges were made Murchison sent him a copy of the revised *Siluria*, "earnestly hoping that the passages relating to yourself in the Preface . . . and the alteration of a phrase or two in the body of the work may remove from your mind the impression produced by perusal of the first edition.

"Time rolls on, and as we passed many a happy day together, I trust you will have some gratification in turning these pages."

That hope was not justified. Sedgwick waited a long time to reply in a stilted note of acceptance which began "Dear Sir Roderick . . ." It showed that the writer had gone to Dent while *Siluria*, seemingly not unwrapped, was left behind in Cambridge.

The conflict meanwhile was being solved with justice to both

The Cambrian Conflict

participants. Sedgwick himself took the first step by spending two summers (1842–43) in an effort to find true Cambrian strata below the Silurian of North Wales. The search was a dismal failure, though it did discover a great gap in the middle of Murchison's system. Not only were Lower, or early, Silurian rocks bent and tilted; they plainly had been crumpled into ancient mountains which then were destroyed by the rains, streams and frosts of a long terrestrial epoch. When the late Silurian sea laid down new strata it spread them upon the roots of that eroded highland.

Sedgwick made little use of this fact, but went off in a desperate search for unquestionably Cambrian fossils. He found them in 1846: a group of small, rather insignificant shells and hardened trails in beds of sandstone. Better luck awaited Joachim Barrande, a French exile in Austria who wrote several large volumes on fossils from both divisions of Murchison's Silurian. After working on these for many years he discovered a third fossil fauna: an assemblage of shelled animals which, for lack of a better term, were called Primordial. Later discoveries showed that many species had relatives in British strata which lay thousands of feet below the lowest supposed Silurian. There plainly was an older system, just where Sedgwick had tried to place it in 1842!

These facts caused hesitation and hedging, for uncompromising stands had been taken by friends of both Sedgwick and Murchison. Slowly, however, the quarrel died down as the Cambrian System was strengthened by new discoveries. Yet there was at least a temporary flare-up when a professor at St. Andrew's, Scotland, restudied fossils from Murchison's Lower Silurian. They were so distinct, he concluded, that this name was deceptive as well as an anachronism. Let questions of credit and friendship be put aside; names were meant not to honor men but to distinguish formations. Since the "Lower Silurian" was not Silurian at all, why not give it a name distinctly its own and on a par with those of other systems? The professor recommended Ordovician, from a tribe which once held the hills in which its rocks were discovered.

The Story of the Great Geologists

Neither Sedgwick nor Murchison lived to see this new term accepted. The former died in 1873, having given up his professorship barely two years before. He already had made a half-hearted peace with Murchison, writing once in 1869 and sending greetings through Lyell a year later. To others he

Sedgwick in academic costume, from a silhouette.

remained as he always had been: a kindly, weather-beaten man who looked almost like a tramp in the field but at college wore the neat black clothes of a clergyman. His lectures were stimulating though discursive; his Elizabethan anecdotes were famous in both Norwich and Cambridge. Now and then he even managed to laugh at his ills in letters to relatives or exceptionally close friends.

Murchison's life, by contrast, was an almost continuous triumph. He still overate and took many pills, yet retained his youthful ability to turn every undertaking into triumph. He could work cautiously, as on the Silurian System, but when a

The Cambrian Conflict

survey of Russia called for speed he toured steppes and forests at a gallop and geologized while his carriage crossed the Urals behind horses sweating under the lash. Returning to London, he kept the Geological Society astir and in time became a catalyst for British science in general. With his help young men secured positions, shy scholars came into official notice, and associations that had been in the doldrums enjoyed crowded meetings. Now and then a critic muttered "meddler" or "politician," but his help was given with such generous good will that objections were very few.

Murchison was knighted in 1846, an honor explaining Sedgwick's chill note with its "Dear Sir Roderick." In 1854 a group of scientific notables urged him to assume directorship of Britain's none too active Geological Survey, along with its School of Mines and Museum of Practical Geology. Within ten weeks he improved all three, felt "in better health" than when he began, and found time to give open-air lectures on geology as well as to carry on his general affairs by means of public and private dinners. Small wonder that his course aroused envy in the gouty, nervous Sedgwick!

Still, success brought penalties. The young Lieutenant Murchison had been conceited; the mature leader of science and social lion devoted more and more thought to his own accomplishments. In time he became impatient with lesser men and unwilling to be corrected on fact or opposed in policy. With these traits came an almost arrogant attitude whose effects were exaggerated by formal manners and military bearing. To some workers that bearing alone carried warning. Take care lest you stand in the way of a man so vigorous and so perennially youthful!

Vigor was there until the end, but youthfulness of mind disappeared in growing conservatism. Progressive workers were amazed to see Murchison drift back to belief in convulsions, abandoning ideas about the erosion of valleys which he and Lyell had developed in 1828. With Von Buch he rebelled against the theory of great glaciations, answering it with denial

The Story of the Great Geologists

and protest rather than argument. He clung to his Lower Silurian, refusing to quarrel only from his love for Sedgwick and because he took little stock in ideas opposed to his own. He could not find "one scintilla of evidence to support Darwin's theory," rejected most of Lyell's revived Huttonism, and sputtered when his own discoveries—and errors—were not mentioned in the *Antiquity of Man*. "As for myself," he wrote, "I am only once mentioned, in order to be knocked down." Yet he surely would have resented further quotation had it led to the same result.

Murchison died in 1871, after eleven months of increasing paralysis. His place on the Survey was taken by a man whom he had trained; one who saw his faults as well as his virtues, yet who argued for "Lower Silurian" when he lectured at Baltimore in 1896. Even now the name sometimes is used in Europe, a faint echo of century-old conflict in which both opposing parties were wrong as well as right.

BENJAMIN SILLIMAN
first professor of chemistry and natural history at Yale.

After Youmans

Merrill

[SAM]UEL L. MITCHILL
[do]ctor, scientist and poli[tic]ian of early New York.

Courtesy Rensselaer Polytechnic Ins

AMOS EATON
at the age of forty.

CHAPTER X

Agassiz of the Ice Age

MUCH OF EUROPE is mantled by a sheet of sand, gravel and clay mixed with boulders, which some people knew as drift and some as diluvium. It was, said Von Buch, the result of upheavals which sent streams of rock rubbish into lowlands and tossed huge stones from hill to hill. Others saw it as one product of Noah's Deluge, which dug valleys, churned soil and huge fragments together, and then dropped them upon the surface as God bade its waters recede. Even Lyell imagined violent local floods and icebergs that floated over lands conveniently submerged to aid in their distribution. Few were prepared and many were indignant when a young Swiss supplanted these notions with fields of slowly moving ice.

Jean Louis Rodolphe Agassiz was born on May 28, 1807, at Motier on the shores of the Lake of Morat, in west central Switzerland. There his parents had moved after years in the bleak village of St. Imier, where the elder Agassiz preached confidence in God to peasants who needed proofs of divine beneficence after labor in their rocky fields. In such surroundings pastor was little better off than peasant and both approached poverty when crops were poor or dairy herds met

The Story of the Great Geologists

disaster. To these hardships was added death, which took child after child till both the pastor and his wife gave up. After praying over a fourth small grave the couple packed their possessions and moved.

Motier was neither large nor rich, but it seemed both to the Rev. Agassiz and his wife. Soil of the farms was deep and good; the parsonage was substantial; its vineyard promised an income in addition to salary. Garden and orchard provided food, while a spring brought plentiful water to a stone pool just behind the house. That pool formed the aquarium in which small Louis kept fish when he learned to catch them in near-by shallows of the lake.

For the boy became a naturalist almost in infancy. He began with rabbits and guinea pigs, but in time reared field mice, snakes and even wild birds in cages under trees of the yard. In fishing he used his hands, treating captives so carefully that they swam and ate vigorously when he transferred them to the stone basin. There the boy sat with his nose close to the water, watching their movements as he noted their shapes, their colors, and the delicate bones in their fins.

Another quality appeared early; that of leadership. Louis guided his playmates and urged them to adventure; his younger brother followed him in both natural history and games. When Louis went fishing so did Auguste; when Louis wanted to skate across the lake Auguste came without hesitation. A crack in the ice seemed too wide to cross, but when Louis turned himself into a bridge his brother crept over it without fear. The mother spied them through a telescope and in terror sent a workman on skates to bring the youngsters home. He led them back across the lake, and Louis was the one who protested. If the ice was unsafe for two small boys, how could it be less dangerous for two boys and a man? And besides, why not go to town and ride home in comfort with Father?

Until Louis was ten he studied at home, learning to read, write and figure with Pastor Agassiz as his teacher. He also

Agassiz of the Ice Age

studied less formally under workmen—the cobbler who came twice a year, the tailor, the carpenter, and the cooper who tightened old casks or made new ones before grapes were ready to harvest. When eight or nine years old the boy could make pens for his pets and could cut and sew leather shoes for Sister Cécile's dolls. He also knew how to mend his own torn clothing and make a model barrel that was watertight. As a man he would credit his skill in dissection to this early training in crafts.

At ten Louis entered the college, or school, for boys at Bienne; Auguste, true to form, followed a year later. Discipline was strict and work was hard: nine hours of study per day, with mathematics, Greek, Latin and modern languages. But there were frequent recesses for games, and home was only twenty miles away. When vacation came the boys made this trip on foot, saving money to buy inexpensive books which Louis always selected. Sometimes they went to Cudrefin instead of Motier for a stay with their mother's father, the old and respected Dr. Mayor. He let them drive his small white horse or come to Easter celebrations where vast numbers of eggs and fritters were eaten and a special dance was announced in honor of the Mayors. Only the family, friends and some neighbors took part while other villagers looked on.

Four years Louis Agassiz spent at Bienne, and then drew up a plan for his future education. It expressed willingness to serve eighteen months of apprenticeship in commerce, after which he was determined to "advance in the sciences" and "become a man of letters." Such a program meant immediate studies in Greek, Latin, Italian, and geography both ancient and modern, with books that would cost twelve louis. Apprenticeship would be followed by four years at a German university, after which final studies in Paris would consume about five years. Then, at the age of twenty-five, he could begin to write.

An ambitious plan for the son of a village pastor who received much of his salary in food. Louis broached it first to

The Story of the Great Geologists

Grandfather Mayor, who was impressed by the boy's notebooks and letters of recommendation from the faculty at Bienne. But there was, he knew, the problem of money; of money to be spent then with care, and of more that must be earned as Louis became a man. Why not take up medicine, which could furnish a reliable income?

Because, began Louis—and then stopped. Medicine, after all, was a science close to zoology and vastly more attractive than business. Grandfather would take his part, and so would Uncle Matthias, a noted physician of Lausanne. There was an evening filled with examination of more notebooks, re-reading of letters, and careful budgeting. In the end it was decided that Louis should continue his college work at Lausanne, with instruction in anatomy under his famous uncle. Since Auguste was too young for business he would go along.

For two years the boys stayed at Lausanne, turning their room into a zoo and delighting a professor who had charge of the canton museum. Then Louis entered the medical school at Zurich, still with Auguste in his wake. The boys lived in a private house, made one inspiring trip to the Alps, and spent many hours copying books which they could not afford to buy. Two years of this and they finally parted, Auguste to enter business while his brother went on to Heidelberg and a medical degree.

Louis Agassiz was not quite nineteen when he entered Germany in April 1826. A pastel sketch shows him as a wavy-haired youth with thin mustache and a wide, open collar. He rose at six o'clock, went to lectures at seven, and kept busy until nine at night. He was disturbed by a charge of six crowns for matriculation, pleased with his new professors, and happy to meet a student of botany named Alexander Braun. They soon became inseparable, and when Agassiz fell ill of typhoid he was taken to the Braun home in Carlsruhe. He left it with a pledge to marry Alexander's sister, Cecile.

Illness was followed by several months at home, during which Agassiz studied the growth of tadpoles, collected fish

Agassiz of the Ice Age

from mountain lakes, and gathered specimens of plants. In October 1827 he joined Braun in a "pilgrimage" to Munich, where lectures were free, board was cheap, and beer good as well as plentiful. On the way they saw their first llama, stuffed and in a museum near the skeleton of a mammoth, which then was called a *carnivorous* elephant.

Pastor Agassiz had described his son as "courageous, industrious, and discreet"; a youth who "pursues honorably and vigorously his aim, namely, the degree of Doctor of Medicine and Surgery." Yet honor did not preclude detours, and by the summer of 1828 Louis was busy writing a book about fishes which his Professor von Martius had brought from Brazil. The work was in Latin, with forty colored plates, and to grace its title page the author became a doctor of philosophy. He received the M.D. degree on April 3, after nine days of examination. To his parents it was a "most precious laurel" which assured their son of a career "as safe as it was honorable."

At twenty-two Louis Agassiz cared nothing for safety, nor did he want a medical career. He had begun a great book on fossil fishes and another describing those of lakes and streams. With an allowance of two hundred and fifty dollars per year he supported himself, traveled to study specimens, and paid an artist to draw them. On borrowed money he reached Paris in 1831, where the foremost scientists entertained him in spite of his poverty. A year later he became professor of natural history in the new college at Neuchâtel, just across the lake from Grandfather Mayor's home. The institution had fewer than a hundred students, little money and no buildings; Agassiz received only eighty louis (about four hundred dollars) per year and lectured in the city hall. He had to start his museum in a refuge for orphans and turn the home to which he brought his bride into a combination of laboratory and boardinghouse in which two assistants, a collector and an errand boy, helped him with his work. The collector brought in fossils which no one else could find, but he also slept in his clothes and seldom changed them. He undoubtedly helped convince the young Mrs. Agassiz that people of Neuchâtel were less pleasant than those of her native Carlsruhe.

The collector's fossils were stuff for another big book, but before it could be written Agassiz found a new enthusiasm. In the summer of 1836 he took his wife and baby son to Bex, where Cecile could relax with Mme. de Charpentier, German wife of an amateur naturalist who managed the local salt works. Charpentier agreed to guide Agassiz on trips through the mountains in search of petrifactions and fish. There also would be visits to gravel ridges and trains of huge boulders supposed to have been left by glaciers that once spread into the central Swiss plain and across the slopes of the Jura.

This supposition was not new with Charpentier; he had received it back in 1815 from a thoughtful mountaineer. For fourteen years the suggestion lay fallow; then it was taken up by a friend who wrote a paper on changing climates of the Alps. Charpentier at last was convinced, as was a learned political refugee from Breslau. But to Agassiz the theory of ancient, long-vanished glaciers seemed nonsense, though he was willing to examine the evidence. Charpentier and he could have grand trips together, even if they found no ancient ice.

Instead of playing the part of skeptic, Agassiz was carried away by the truth of his host's "nonsense." Together they examined the existing glaciers of Diablerets and Chamonix, where Lyell had worked, followed ridges of drift along the Rhone, and traced those of tributary valleys. Water could not have piled up such deposits; only moving, melting ice could have left them after scouring and scraping bedrock, at the same time carrying great boulders scores of miles from their source. And if glaciers had done such work near the Rhone, must they not have operated wherever drift, scoured bedrock and "erratic" boulders were found? Agassiz insisted that Charpentier publish his facts, in the hope of stimulating other naturalists to make similar observations. He himself went glacier-hunting in the Jura and, when the Helvetic Association

Agassiz of the Ice Age

met in 1837, was ready with a presidential address which one critic called a "fiery discourse about a sheet of ice."

One rare quality of Agassiz was his power to see great subjects as wholes; to state large problems and conclusions so comprehensively that his first work commanded respect. He had done this in the generalizations of his preface to *Fossil Fishes;* he now displayed his ability as well as daring by going directly from the evidence for ancient, long-melted glaciers to the fact of a glacial period. That period, he suggested, had begun with a temporary but world-wide and rapid change in climate which allowed a sheet of ice to spread from the North Pole to central Europe and Asia. "Siberian winter," he concluded, "established itself for a time over a world previously covered with a rich vegetation and peopled with large mammalia, similar to those now inhabiting the warm regions of India and Africa. Death enveloped all nature in a shroud, and the cold, having reached its highest degree, gave to this mass of ice, at the maximum of tension, the greatest possible hardness."

Von Buch was in the audience; his disapproval, tinged with tolerant contempt, bubbled over when the young president stopped speaking. Humboldt, a stanch friend and benefactor, expressed his objections by letter. "Over your and Charpentier's moraines Leopold von Buch rages, as you may already know. . . . I, too, though by no means so bitterly opposed to new views, and ready to believe that the boulders have not all been moved by the same means, am yet inclined to think the moraines due to more local causes." And again, "Your ice frightens me. . . . I am afraid you spread your intellect over too many subjects at once."

Agassiz did not share that fear, nor was he alarmed by ice. Besides his work as professor he was printing and illustrating *Fossil Fishes* and *Fresh-Water Fishes,* as well as books on mollusks and on the group to which sea urchins belonged. He maintained a lithographic plant with as many as twenty workmen, publishing books by other authors to keep his employees busy. Yet when August of 1838 came round he was ready for a trip to the glaciers of Mont Blanc, and a year later studied ice streams of Monte Rosa, the Matterhorn, and several other mountains. At one place he examined a cabin built on ice in 1827; in twelve years it had traveled four thousand feet and seemed to be gathering speed.

In 1840 Agassiz published his *Studies on Glaciers,* with one large volume of text and an atlas of thirty-two plates. It reviewed earlier work on the subject and presented new facts about the appearance and structure of these streams of slowly moving ice, their formation and internal temperatures, and the loads of broken and pulverized rock scattered through them or carried on their surface. Most stimulating, however, were chapters dealing with ancient Swiss glaciers and with the ice sheet which had brought arctic conditions to a once-warm continent.

The surface of Europe [he wrote], adorned before by a tropical vegetation and inhabited by troops of large elephants, enormous hippopotami, and gigantic carnivora, was suddenly buried under a vast mantle of ice, covering alike plains, lakes, seas and plateaus. Upon the life and movement of a powerful creation fell the silence of death. Springs paused, rivers ceased to flow, the rays of the sun, rising upon this frozen shore (if, indeed, it was reached by them), were met only by the breath of the winter from the north and the thunders of the crevasses as they opened across the surface of this icy sea.

The main case was presented; now for details. In the summer of 1840 Agassiz hired a mason to wall in a huge block of micaceous schist that stood on a ridge of rock rubbish, or moraine, bisecting the lower Aar glacier. The uneven floor was smoothed with slabs, a blanket served as door, and the whole formed a hut that would shelter six people if none minded his neighbor's elbows. A niche under another boulder served as storehouse for food.

To this hut, the "Hôtel des Neuchâtelois," Agassiz took friends and guides for detailed investigation of the glacier. They carried heavy loads of instruments, as well as an auger

Agassiz of the Ice Age

for boring holes through which recording thermometers could be lowered into the ice. Microscopes were ready for examination of insects, small plants and other things living upon the moraine. An engineer determined the position of eighteen large boulders which were to be sighted and plotted year after year to determine how fast the ice was moving.

Close-range studies were varied by longer trips on which the whole party climbed such "unscalable" peaks as the Jungfrau and Schreckhorn. Sometimes they went lightly laden; more often their rucksacks were filled with barometers, thermometers and instruments for simple surveying. On one trip they carried heavy loads through loose snow that came to their knees and then climbed a flight of steps which their guides chopped in the ice. At the summit they began a peasant dance, to stop abruptly as a band of chamois appeared from behind a rock. The trip down was largely a matter of sliding, though they traversed one great crevasse on an ice bridge one to two feet in width and broken near the end. There, one of the climbers noted, "we were obliged to spring across."

Results of this work appeared in the *Glacial System*, published in 1846. Meanwhile Agassiz had been to Britain, where Buckland (Lyell's old teacher) had forsaken deluges for ice. In 1840 Lyell himself was convinced, finding a chain of Ice Age moraines two miles from his father's house, Kinnordy. He and Buckland supported Agassiz at a meeting of the Geological Society of London in November 1840, and though Murchison tried to oppose them, his arguments had little weight. Two years later, Charles Darwin traced ancient glaciers of North Wales and was happy to find a moraine which Buckland had overlooked. The King of Prussia granted almost a thousand dollars for further work on the Aar, and Agassiz began to plan a trip to North America. But he delayed it to complete the *Researches on Fossil Fishes* as well as a volume on queer fishlike creatures from Devonian sandstones of the British Isles. He also hastened to complete other works, to settle affairs of the college at Neuchâtel, and to close

The Story of the Great Geologists

his profitless printing and lithographing plant. No wonder that Humboldt wrote him: "For pity's sake husband your strength!"

At two o'clock on a night in March 1846 Agassiz left Neuchâtel for Paris and the United States. The King had given him fifteen thousand francs for travel; John A. Lowell had engaged him to lecture at the Lowell Institute in Boston. On shipboard Agassiz practiced English; on landing he hastened to survey the country between Boston and Washington. To his mother he wrote enthusiastically about the "frightful" speed of American trains, the hospitality of cultured people, and the boulders, moraines and ice-scoured bedrock of eastern Massachusetts. He also reported the "conversion" of several scientists whose ideas about drift had ranged from hazy to grotesque.

Agassiz gave his double course of English lectures, which were free, and added a subscription series in French which prosperous young ladies of Boston attended with enthusiasm. He left the young ladies to study turtles near Charleston, where jellyfish also were most attractive. By April he was back in New England, leasing a house in East Boston at a high rental partly because it was large and partly because it had a back yard that dipped into the harbor. In this house Agassiz welcomed assistants who had been with him on the Aar and at Neuchâtel, as well as the pastor—now an exile—who had paid for his first trip to Paris. Papa Christinat managed the household while Agassiz went to Niagara Falls, collected fish along the St. Lawrence, and lectured in every large city from Albany back to Charleston.

Agassiz had planned to go back to Switzerland after three years of travel in the United States. But the spring of 1848 found Europe in turmoil; French revolutionaries had proclaimed a republic, Neuchâtel had rebelled against Prussian rule, and several friends besides Papa Christinat were exiled from their homes. Agassiz had no wish to return, but when his Prussian travel grant stopped he had to hunt a job. He was

Agassiz of the Ice Age

happy to become professor of natural history in the new scientific school at Harvard, with an assured salary of fifteen hundred dollars per year. This sum, guaranteed by the founder, must do until students' fees reached three thousand dollars.

The first course which Agassiz gave at Harvard began in April 1848. At about that time he moved from Boston to Oxford Street, in Cambridge, where there was space for a garden and for living animals. Papa Christinat preferred the former, sighing as space was given to useless turtles and a tank for alligators. A tame bear was chained in one part of the yard; eagles had a cage built close to the wall; smaller cages held a family of opossums and a hutch of rabbits to be used in experiments. There also were vast collections of dead things in a shanty perched on piles above the Charles River.

The collections contained few fossils or rocks, for Agassiz was enthralled by the wealth of undescribed animals to be found in North America and surrounding seas. He studied mammals, turtles, fish and odd invertebrates that peopled rocky New England coasts. His kindliness and enthusiasm fired pupils, who turned the new house into a laboratory almost as crowded as the one back in Neuchâtel.

Still, the Ice Age could not be forgotten on a continent where ancient glaciers had covered five million square miles. In the summer of 1848 Agassiz took students to Lake Superior, where they traveled in bark canoes through what then was wilderness. As he had done in Switzerland and Great Britain, the professor showed that currents of water could not have scattered boulders without sweeping far beyond the limits in which they were found. Icebergs were inadequate; moreover, they called for a climate cold enough to produce great icecaps. Scratched boulders and bedrock recalled the Alps; ridges and sheets of drift matched those left as Swiss glaciers melted. Bedded, or stratified, "drift" did not count, for it belonged to a later epoch when lakes and even sea water had invaded the continent.

Cecile Agassiz had never been reconciled to her husband's

impetuous pursuit of science and his unpredictable ways. When he left Neuchâtel she took refuge with her parents in Carlsruhe, where the Agassiz youngsters were happy. Their mother brooded and soon became ill, an illness that ended in death while he was on the trip to Lake Superior. Returning to Cambridge, he assuaged his grief by planning how the children should come to America.

Papa Christinat listened and then attacked. Come to America, indeed! Let Louis look and use his head. Was this a place to bring children—this place which was part boardinghouse, part laboratory, with men squabbling in every room and a zoo in the back yard? Get rid of those quarrelsome assistants; marry some rich American woman—he knew some who already were preening to catch the professor's eye. Achieve security and a civilized home before sending for the children!

Agassiz could have used wealth, for no scientist ever had more need for money or made fewer efforts to get it. But to marry wealth—that was something else; something not to be done by a man whose ways were as unconventional as those of Agassiz. He shook his head at advice, sent for his son, and paid court to Elizabeth Carey. She was a professor's sister-in-law and poor, but what matter? Did she not know how to be companion as well as helpmate of a genius? Was she not eager to mother the girls, and did she not capture young Alexander's heart the day he arrived in Cambridge?

Papa Christinat grumbled and moved; Elizabeth and Louis sent for the girls; the bride turned the dirty, orderless Oxford Street house into a cheerful home. By August of 1850 the Agassiz family was united, although burdened with debts that dated back to the frenzied years at Neuchâtel. Elizabeth pinched pennies for five ineffectual years and then, with Alexander's aid, set up a school for girls on the upper floors of the house. She charged high fees but gave thorough instruction, with the advantage of occasional lessons by Harvard's most famous professor. In eight more years all debts were paid and the school could be discontinued.

Agassiz of the Ice Age

By that time Agassiz was receiving a worth-while salary, a legacy and state funds had given him a museum, and rich men had made grants for research. In 1865 one wealthy admirer sent the professor, his wife, and several assistants on an expedition to Brazil. They collected fourteen hundred new species of fish, while Agassiz mistakenly reported that glaciers had once advanced to the region of Rio de Janeiro. Here he was misled by weathering that had rounded great blocks of stone where they stood and had smoothed bare hills till they looked like the "sheep rocks" of New England or Switzerland.

Better work was done in 1868, when he crossed wide moraines of the prairies and examined the ice-gouged Finger Lakes of New York. In 1872 he and Mrs. Agassiz sailed around Cape Horn and to San Francisco on a vessel of the Coast Survey. Near the Straits of Magellan they found thick moraines, traced the movements of vanished ice, and examined existing glaciers. Though not so thick as ice streams of Switzerland, some of these were much wider than any on the continent of Europe.

This trip almost completed Agassiz's work; after a school year and one more summer of teaching he died in 1873. Some conservatives still denied his great Ice Age; other men were tracing sheets of drift as well as intervening deposits which would show that the Glacial Period consisted of several epochs. Agassiz died before these men published. He thus missed the pioneer's greatest reward, that of seeing a new generation so refine his discoveries that they seem crude and inadequate.

CHAPTER XI

New Science to New World

AGASSIZ WAS by no means the first European who brought earth science to North America. During the Revolution, Dr. Johann Schöpf had come as a surgeon with the Hessians, remaining to tour the East and Southeast after the peace of 1783. He described the "mineralogy" of mountains and farming districts, recognized what we now call the Coastal Plain, and told how it was marked off from hilly regions to the northwest by a chain of steep slopes over which rivers plunged in waterfalls. Thus he recognized the Fall Line, a feature which would control American industry in the period of water power.

Schöpf had been gone for a dozen years when the learned Constantin, Comte de Volney, sailed from Havre for the United States. Volney was a traveler and historian who had lived in Egypt and Syria, had sat in the States-General of France, and had served as professor at the new École Normale. He also had been imprisoned by the Jacobins and was fleeing from the Directory, "whose axe was continually falling on the necks of those whose conduct and opinions" resembled his own. "Saddened by the past and anxious for the future," Volney wrote, "I set out for a land of freedom, to discover whether liberty, which was banished from Europe, had really found a place of refuge in any other part of the world.

"In this frame of mind," he continued, "I visited almost every part of the United States . . . and such was the contrast which the scene before me bore to that which I had left, that I resolved to make it my future residence." But in the spring of 1798 a "violent animosity" broke out against France, giving rise to suspicions that were not allayed by the friendship of respected men, including Washington. Volney found himself condemned as a spy and conspirator who was plotting to seize Louisiana for the Directory. The would-be immigrant had to give up his plans and return to France, where he received protection and some favor from Napoleon. In 1803 he published a volume called *A View of the Soil and Climate of the United States of America*. British and American translations appeared a year later—proof that the work was influential and that it filled a need.

Volney was not a geologist, and his discussion of land merely provided background for longer and more ambitious chapters about climate, diseases, history and aboriginal Indians. After a general geographic introduction he outlined and described five regions characterized by different rocks and contrasting structure. The first of these, he thought, was dominated by granite; it stretched from Long Island to the Gulf of St. Lawrence, westward as far as Kingston and southward again to New York. "Calcareous masses" were scattered among the granite, though directly imbedded in platy rocks which were lumped together as schist.

"The grit or sandstone of the Katskill," wrote Volney, characterized a second region which included the mountainous Blue Ridge, the Allegheny country, and highlands as far south as Georgia. He "lost track of it in the state of Tennessee," but thought that the sandstone merged into a great central basin underlain by level strata of limestone and covered with rich black soil. This limestone region extended westward at least to the Mississippi and much farther into the Northwest.

The Story of the Great Geologists

Volney collected fossils at Onondaga, New York, and Cincinnati, taking them home to France. There the pioneer evolutionist, Lamarck, "assigned them to the *terrebratula* genus and declared that strata from which they came had once been the *bottom,* and not the *banks* of the sea." Remains of shore-dwelling creatures, he rashly concluded, were lacking from North America.

No fossils were mentioned from the Coastal Plain, which Volney called the region of sea sand. Its western boundary was a bank of "granitic talc, glimmer, or isinglass"—in modern terms, the eastern Piedmont with its schists, gneisses and granitic rocks. Volney clearly described the Fall Line, as well as sands, clays and beds of muck in which plant remains were abundant. But he had no conception of the tilting and overlapping which carry successive formations underground between the Fall Line and the sea.

Volney's fourth region, that of river-formed soil, was the Piedmont itself. Lying between mountains and the Fall Line, it was crossed by streams whose flow supposedly had been torrential when the Appalachians were young. On the lowlands their silt-laden waters had spread, leaving layer on layer of sediment which in time covered older rocks such as the "granitic talc."

Though Volney dwelt upon the vigor of streams, he doubted that they could have worn the precipitous water gaps through which they emerged from the mountains. "The more I consider the situation . . . ," he wrote, "the more I am confirmed in the belief that the Blue Ridge was once entire," damming the ancient Potomac and Shenandoah to form enormous lakes. Other ranges ponded additional streams, whose waters finally spilled over low points in the barrier. Gaining force, they began to cut and scour, making their channels wider and deeper until every basin was drained.

Though many fossils from the limestone region were marine, Volney thought that it too gave evidence of long-vanished lakes. Such stagnant waters, he said, "account for the levelling of the earth throughout the western country in horizontal strata; they tell why these strata underlie each other in the order of their specific gravities and why, in many places, there appear the remains of trees and other vegetables, and even of animals, such as the bones of the mammoth ... which could only have been thus collected by the action of

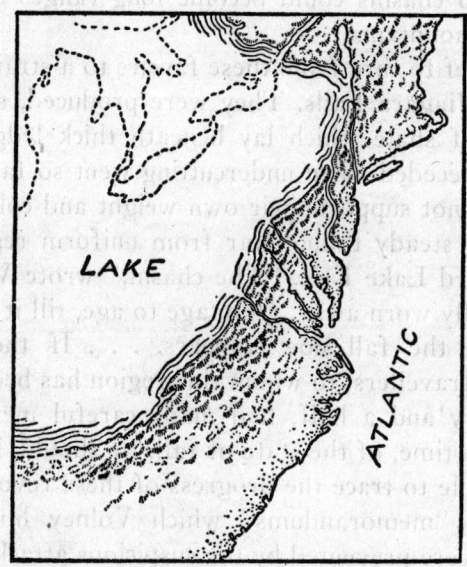

Map illustrating Volney's theory of a great ancient lake which overflowed in streams that cut gaps through the Appalachian Mountains.

water. In short, a happy and natural explanation is thus offered for the formation of those deposits of coal which predominate in certain situations and districts."

Coal Age plants and Ice Age mammoths in one lake; the mixture shows how poor was Volney's conception of earth history. Nor did he do better with earthquakes, which were said to be common in the eastern United States. They were caused, he thought, by a line of subterranean fires which ran northwestward through a "schistous stratum" that extended to

The Story of the Great Geologists

Lake Ontario. These fires had caused countless great explosions which allowed granites, limestones and other rocks to be shattered and tumbled into chasms. Lake Ontario itself was the crater of a huge volcano—a conclusion strengthened by the many volcanic substances found upon its shores. What these substances were Volney did not say, nor did he show how rocks that fell into chasms could become long ranges of crumpled and broken mountains.

It is a relief to turn from these fancies to a straightforward account of Niagara Falls. They were produced, said Volney, by erosion of shale which lay beneath thick ledges of limestone; they receded when undercutting went so far that those ledges could not support their own weight and collapsed. The result was a steady though far from uniform regression upstream toward Lake Erie. "The chasm," wrote Volney, "has been gradually worn away, from age to age, till it reached the place where the fall now appears. . . . If the European colonists or travellers, to whom this region has been accessible for a century and a half, had made careful memorandums, from time to time, of the state of fall, we should, by this time, have been able to trace the progress of these revolutions."

Any such "memorandums" which Volney himself might have made were prevented by the suspicious attacks that burst on him in 1798. For five years little or nothing was done—not merely about Niagara Falls, but upon formations and mineral resources in North America. Then work was begun by a wealthy Scot, who brought a Wernerian type of geology from Europe and applied it in the New World.

William Maclure was a mixture of shrewd businessman, sanguine reformer, philanthropist, and man of science. Born at Ayr, Scotland, in 1763, he was privately taught by a Mr. Douglass, who enjoyed fame as a pedagogue in classics and mathematics. But an adventurous enthusiasm for business prevented much formal education, for when only nineteen young Maclure sailed to New York in search of some profitable connection. Making the "requisite arrangements for mercantile

New Science to New World

employment" without delay, he took the first ship back to England, there to become partner in an exporting and importing firm whose offices were in London. Both the firm and Maclure himself were successful, for he built up a fortune which would allow him to retire while still a young and vigorous man.

Maclure returned to the United States on business in 1796, found the nation better for seventeen years of progress, and decided to remain. When he journeyed to Europe again in 1803 he did so as one of three commissioners sent to settle claims of Americans whose property had been destroyed during the French Revolution. With that business done he went on trips from Britain to Bohemia and from Italy to the Baltic; trips on which he gathered specimens and bought books for shipment across the Atlantic. He also conferred with scientists, whose work was not much disturbed by Europe's recurrent wars.

Maclure came back to the United States full of enthusiasm for science, and especially for geology as an element in education. Why, he asked, should young people of the United States spend their time learning Latin, Greek and other time-honored subjects? They would do much better to study the earth and its resources, gathering useful information while they trained their minds. For geologic facts were needed before Americans could develop their supplies of iron, copper, zinc, coal and clay, while an understanding of soils was needed to properly manage farms. Surely it would do more good than declension of Latin nouns!

Education, of course, presupposed information, yet very few data existed on American ores and rocks. Nor were bureaus, surveys or consultants at hand to secure it; if Maclure wanted geologic facts he must get them for himself. He therefore went forth, wrote a contemporary, "with his hammer in hand and wallet on his shoulder, pursuing his researches in every direction, often amid pathless tracts and dreary solitudes, until he had crossed and recrossed the Allegheny Mountains no less than fifty times. He encountered all the privations of hunger, thirst, fatigue and exposure, month after month and year after year, until his indomitable spirit had conquered every difficulty and crowned his enterprise with success."

This account was much too enthusiastic; Maclure neither made a thorough survey of Eastern formations nor one of great practical value. But in 1809 he did publish a booklet of *Observations on the Geology of the United States,* accompanied by a colored map. It was the first truly geologic map showing any part of North America, and one of the earliest in the world.

In Europe, Maclure had been deeply impressed by German science and had regarded Werner's classification as the "most perfect and extensive in its general outlines" of any then available. But when he came to discuss American rocks he violated Werner's rules by distributing trap, limestone and gypsum throughout the Primitive, Transition and Secondary "classes." He also made an effort to distinguish between older and younger deposits of each kind within one class, listing three Flötz sandstone formations of varying ages, as well as two distinct Flötz "traps." Under these he grouped a variety of dark rocks including basalts and columnar diabase in the Palisades of the Hudson.

Maclure continued his studies in the United States after 1809 and also visited twenty islands of the West Indies. By 1817 he was able to republish his *Observations,* correcting geographic errors and adding chapters on rock decomposition and soil. But he was still enough of a Wernerian to give only limited credence to the work of William Smith.

In the alluvial of New Jersey [he wrote], about ten to twenty feet under the surface, there is a kind of greenish blue marl, which they use as manure, in which they find shells, as the Ammonite, Belemnite, Ovulite, Cama, Ostrea, Terebratula, Etc. Most of these shells, are similar to those found in the limestone and grey wacke of the transition, and equally resemble those found in such abundance in the secondary horizontal limestone and sandstone; from which it would follow, that the different classes of rocks on the continent cannot be distinguished by their shells,

New Science to New World

though the different strata of the same class may be discovered and known by the arrangement of the shells found in them.

In 1819 Maclure went to Spain, where a liberal constitution had been proclaimed. He bought some ten thousand acres of land, repaired the buildings, and was ready to open an agricultural school for poor farmers when the new government was overthrown. Priests laid claim to his land, once seized from the Church, and Maclure found himself in danger of capture and enslavement by brigands, who then would have held him for ransom. Since the government was not much better than the robbers, Maclure slipped out of Spain and returned to the United States.

He still dreamed of an agricultural school and of social reform which would make men value learning and useful work rather than position or riches. Both seemed possible at New Harmony, Indiana, where Robert Owen was spending thousands of pounds to found a communistic colony along the banks of the Wabash. Maclure interested several scientists in the project, established his school of agriculture, and moved his library and collections from Philadelphia. When the colony crumbled he persevered until failing health forced him to seek relief in Mexico. There he laid plans to help Indians by giving able young red men training in the United States. He died at San Angel, southwest of Mexico City, in March of 1840.

Maclure was a man of affairs who accomplished more for science with his money than by his own research. He helped establish the Academy of Natural Sciences, in Philadelphia, gave it some thirty-three hundred books, and endowed it with twenty thousand dollars. He brought scientists to the United States, imported the plates of foreign books, and sponsored American editions to be sold at less than cost. He was one of the founders of the American Geological Society and its president until his death. That it failed soon afterward gives some measure of Maclure's influence while he lived.

Some beginnings in earth science, however, were made by

native Americans. Memoirs of the American Academy of Sciences for 1785 contained papers on oilstones, minerals and fossil shells. Two authors regarded iron ores of Connecticut as lavas of a volcano that had erupted twenty-seven years before, with flames and a "louder noise than common." In 1793 Benjamin De Witt described a variety of stones and limy petrifactions in the drift near Lake Ontario. They were, he thought, the products of "some mighty convulsion of nature, such as an earthquake or eruption." The lake itself might be "one of those great fountains of the deep which were broken up when our earth was deluged with water."

Better work was done by Samuel Mitchill, a doctor, teacher and legislator whose home was in New York. Mitchill had studied both medicine and law, and in 1792 became professor of chemistry and natural history in Columbia College. He already had helped found a Society for the Promotion of Agriculture, Manufactures and Useful Arts, with the object of finding coal in New York State. As the society's commissioner, Mitchill produced a *Sketch of the Mineralogical History of New York* which was published during 1798 and 1801 in his own magazine, *The Medical Repository*. He divided the mainland of the state into regions of granite, slate, limestone, sandstone and alluvium. Slate, he said, overlapped granite on the north and west; limestone was distributed over both granite and schist; sandstone extended westward from the "Kaatskills." Alluvium was widely scattered, since it filled intermont valleys and followed the course of streams.

Some of this suggests Volney, and without doubt the two men did exchange or compare ideas. A still closer resemblance appeared in 1818, when Mitchill added his *Observations on the Geology of North America* to the New York edition of Cuvier's *Essay on the Theory of the Earth*.

Cuvier was a German who dominated French science; a politician and administrator who also was the greatest anatomist of his time, an authority on ancient mammals, and an opponent of evolution. The *Essay* had originally appeared as the

New Science to New World

introduction to a great work on fossil bones—an introduction which advocated catastrophes as the cause of extinction, with emphasis on sinkings and elevations which were "instantaneous instead of gradual . . . sudden and violent." In these swift changes of land and sea "living things without number were swept out of existence. . . . Those inhabiting the dry lands were engulfed by floods; others whose home was in water perished when the sea bottom suddenly became dry land; whole races were extinguished, leaving mere traces of their existence—traces which are now difficult of recognition, even by the naturalist."

Mitchill accepted Cuvier's ideas without reservation, but lacked data on which to apply them widely in North America. His principal concern was to describe a series of inland seas whose eastern limit was a barrier formed by the Adirondacks, Catskills and Blue Ridge. In time the wall broke near the present outlet of Lake Ontario and at other points as far south as Harpers Ferry. The Thousand Islands were "scattered monuments of the ruin" left by waters rushing seaward through the St. Lawrence Valley, while water gaps of the Delaware, Susquehanna and Potomac were outlets scoured with such violence that no rocky "monuments" could survive.

Little science was taught in American colleges of the early 1800s, largely because both clergy and important laymen clung fiercely to a literal Genesis. Columbia had braved their opposition, however, and in 1802 Yale followed with a chair of chemistry and natural history. Since no qualified foreigner might be hired, the appointment went to a youthful lawyer who agreed to fit himself for his job.

Benjamin Silliman then was twenty-two; a Yale graduate who had convinced himself that the study of nature was "elevated and virtuous, pointed toward the infinite Creator." As soon as his professorship was assured he hurried to Philadelphia, which then was the scientific capital of the United States. There he studied chemistry at the Medical College, anatomy with Dr. Casper Wistar, and botany under B. S. Barton. After two winters of such training Silliman returned to Yale and wrote a series of sixty lectures which "made some mention of mineralogy." His first class included John C. Calhoun, John Pierpont and the future Bishop Gadsden, of North Carolina.

As soon as the money could be spared Yale voted ten thousand dollars for scientific books and apparatus, all to be bought in Europe. Silliman therefore went abroad in the dual role of purchasing agent and professor on sabbatical leave. He bought books, chose equipment, met scientists and, especially in Edinburgh, attended lecture courses. Jameson had not begun to teach, but John Murray mixed impassioned Wernerian harangues with discourses on chemistry. He was answered by a Dr. Hope, who with equal vigor defended a Vulcanism more sweeping than that of Hutton.

Silliman absorbed the ideas of both schools and came home determined to try them out on rocks of Connecticut. In 1810 he published a *Sketch of the Mineralogy of the Town of New-Haven,* which laboriously proved that most of its ground was an alluvial plain. Silliman described hills of basalt that lay upon sandstone but said that East Rock was made of once-molten stuff that had cooled and hardened underground. A near-by rock was much more puzzling, since he had to pronounce it granite, "though it is not a granite, and inclined to whin, although it is not a whinstone."

Chemist thus grew into geologist, and then into public speaker, editor and general factotum of science. In 1808 Silliman gave a course of popular lectures that led to several "happy unions," including his own marriage to Governor Trumbull's daughter. In 1834 he began to travel, lecturing in Hartford, Boston, Mobile, Natchez, and many other cities. Twelve hundred people paid to hear him in Boston; in New Orleans the *Times-Picayune* announced that "Professor Silliman's introductory lecture was attended by one of the largest and most intelligent audiences ever convened in this city. . . . We predict that these lectures will prove in a high degree in-

New Science to New World

structive and entertaining, and one of the most gratifying sources of popular entertainment."

Only St. Louis was at all hostile. There the Christian Association engaged a room in the building where Silliman was to speak, and a bishop tried to forestall him by discussing the "Assumptions of Geology." Half the geologists, he declared, were infidels; the other half were faithful but deluded and therefore deserved pity. Instead of putting Silliman in either group, he let the audience draw its own conclusions. "Rude and uncandid of him!" commented the professor.

Meanwhile the country's first mineralogical journal had died after publishing a few issues. Friends urged Silliman to fill the gap by editing a magazine that would "embrace the circle of the physical sciences, with their applications to the arts, and to every useful purpose." Despite doubts of financial success he agreed and in January 1818 sent out an optimistic prospectus. Some three hundred subscriptions trickled in—enough to let the *American Journal of Science* appear in July.

The new magazine was a success faced with almost immediate failure. Critics praised and authors sent manuscripts, but the rivulet of subscriptions increased only to three hundred and fifty. The first publishers took their loss and gave up; Silliman secured others only by getting bank loans to pay for paper, printing and engraving. Not till 1822 could he report, first that deficits were not increasing, then that "pecuniary patronage" was making the *Journal* "no longer a hazardous enterprise." He weathered a second crisis by becoming full owner and in 1829 valued that ownership at three thousand dollars. At the same time he had to explain that payment could not be made for all copy. "It is earnestly requested," Silliman continued, "that those gentlemen, who, from other motives, are still willing to write for this Journal, should continue to favor it with their communications."

They did so with such willingness that the *Journal* was able to withdraw first from the field of agriculture and then from the arts, both of which had been included in the original plan.

The Story of the Great Geologists

It became a purely technical magazine devoted to chemistry, earth science and the study of fossils. During this change it went from one volume to two per year and from four to six, and finally twelve issues. Today it stands as North America's oldest independent journal of science and one of the foremost in the world.

One man could not keep pace with such growth; Silliman called on his son, his son-in-law, and finally a notable group of assistant editors who gave the magazine much of its character and scientific standing. Believing that young men made good teachers as well as editors, he also asked relief from his professorship in 1849. Yale persuaded him to teach four years longer, after which he traveled in Europe, gave a few lectures, and spent much time writing his memoirs. He died in 1861, at the age of eighty-five.

CHAPTER XII

Ex-Convict Professor

FEW GEOLOGISTS started out in that science early in the 1800s. Lyell and Silliman had been lawyers, Agassiz was a zoologist, Sedgwick a tutor and clergyman, Murchison a wealthy idler who entered science through chemistry. The man who became America's first great teacher of geology did so in spite of detours as attorney, land agent and "lifer" convicted of forgery.

Amos Eaton was born on May 17, 1776, in what now is Chatham, New York. His father was away at war, acting as forage sergeant for a company of volunteers. His mother had returned to her family home, now headed by a stepfather who also was a soldier. The baby was named for Grandfather Hurd, who had starved to death in the French and Indian War.

When Amos Eaton was fourteen he did small jobs of surveying with a homemade chain and compass. From 1791 to 1795 he studied Latin, Greek, geography, logic, natural philosophy and mathematics, preparing himself so thoroughly that he was able to enter Williams College in September of the latter year. He withdrew as a junior in 1797, having passed

[137]

The Story of the Great Geologists

"with reputation" the public examination of his class. It seems, however, that he continued work under a tutor, for he graduated from Williams as a bachelor of arts in 1799. A few months later he married Polly Thomas, the sister of his best friend.

Now came three years of "reading law," first in Spencertown and then in New York City. Eaton was charmed by the metropolis of sixty thousand people, where he became friendly with Washington Irving, was entertained by Samuel Mitchill, and was able to study such subjects as chemistry, botany and the miscellany of physical sciences then called natural philosophy. He was admitted to the bar in 1802, six weeks after Polly had died of "consumption."

Those were days when New York State had many vast properties whose owners employed agents to act as manager, rent collector, salesman, attorney and all-round supervisor. Eaton took such a job under John Livingston, a land-poor and money-grabbing curmudgeon who paid less than six hundred dollars for eight months of labor. When told that this was not enough he burst out with ill-tempered advice on marriage and limitation of families. "You have not a proper Idea of the expence of a large family," he wrote Eaton, then a widower with one three-year-old son who needed mothering. "The one you have you are not in circumstances to support this is absolutely the case, and so I told you—if you are displeased at this—go on and even increase it—but do not do it at my expence."

Eaton answered by marrying Sally Cady, a one-time sweetheart whom the dying Polly had made him promise to court. In 1804 he and his father bought about five thousand acres of land near Catskill from Nathaniel Pendleton, of New York. Amos moved to a village near Catskill, became agent for a quarter-million additional acres, and soon achieved such a position that he could call himself rich.

All went well till September 1809, when a scalawag and a vengeful ex-tenant had Eaton indicted for forgery. The case was quashed in December, but it seemingly put ideas in the

Ex-Convict Professor

head of that greater and much more able scoundrel, Nathaniel Pendleton. In an effort to regain his land (now one third paid for) and drive Eaton far from New York, Pendleton made a second charge in the fall of 1810.

The case was long and complicated; it reads like nonsense unless one knows the laws that burdened New York more than a century ago. But the record shows clearly that Pendleton, a known swindler, first legally robbed Eaton of land to which he had honest claim. That done, Pendleton, masquerading as "the People," had Eaton tried on evidence secured and in part manufactured five years after the first supposed fraud had taken place. Finally, Eaton was indicted, in January 1811, for a specific crime which he supposedly committed 145 days later.

Nonsense? The story would have been absurd had not a jury solemnly found "the prisoner at the Bar Guilty of the forgery whereof he stands indicted." He thereupon was "sentenced to be confin'd in the State prison at hard Labour for and during his natural Life."

The prison to which Eaton was taken stood near the bank of the Hudson in what then was the fashionable town of Greenwich. Local people were proud of the Doric columns and handsome cupola, though the former must have been partly hidden by a wall fourteen feet high in front and twenty-three along the river.

Inmates of the classic structure were supposed to be reformed by labor—twelve hours of it every day. But by 1814 Eaton became a privileged clerk who handled weekly accounts and was allowed any time he might save by working rapidly. He "had the same living as the keepers," walked in the yard when he was tired, and sometimes strolled beside the river with guards on Saturday afternoon. More important, he might sit till bedtime in the agent's back office instead of being locked in a dark and unheated cell. Thus he was able to work on a botanical book, prepare a manual of practical surveying, and copy a system of mineralogy from a book by Hutton's critic, Kirwan.

The Story of the Great Geologists

All this might have meant little had not Eaton become intimate with influential people. John Torrey, Sr., was the prison agent; his son, the future botanist, became a fast friend and admirer. One of Eaton's law teachers was recorder and De Witt Clinton, then mayor of New York, frequently visited the prison. Though evidence is lacking, it seems that these men and Mitchill virtually forced the governor to grant Eaton a conditional pardon in November of 1815. Clinton himself made it "full and absolute" when he became governor less than two years later.

When he was freed Eaton took a steamer from New York, seeing "nothing worthy of attention, but basalt columns 20 miles above the city, on the Jersey shore." He reached Hudson at noon on December 3 and studied slate, hornstone and trap as he walked on toward Chatham. There he visited his family, hunted rocks, and made ready to study them at Yale. Urged on by the limitations of his pardon, he reached New Haven on February 23 and found himself a welcome student. He attended Silliman's lectures on chemistry every morning, with mineralogy—which still meant geology—on Wednesday and Saturday afternoons. Eaton also found time to study plants under Professor Ives, and in mid-March advertised "Proposals for publishing a work, to be entitled, *A Practical Digest of American Botany*. . . . The materials to be selected from books (mostly in Latin) which could not be purchased for one hundred dollars."

Printing evidently was not warranted, for Eaton revised his plan on a more modest scale. Meanwhile he faced—and surmounted—a new disaster. His wife, Sally, had suffered much while he was in prison, for her parents did not like Amos and took every chance to attack him. She was ill when he had to leave for Yale, soon "fell into a state of total derangement," and died suddenly, leaving four small sons. Determined to hold his regained family together, Eaton frankly began to hunt for a wife and married four months later. There was urgency, calculation and no time for love, yet the union was

Ex-Convict Professor

successful. Both parents were delighted when a daughter was born in August of 1817.

By that time Eaton had finished his work at Yale, which then had no facilities for research or advanced study. In March he went to Williams College, where the faculty engaged him to lecture on mineralogy and botany. To these he added geology proper, giving six "extempore" classroom lectures every week while a seventh, written out in full, was presented in the college chapel. He found the professors of language and mathematics the "most persevering and zealous naturalists in New England"; with their help his work became so successful that "an uncontrollable enthusiasm for natural history took possession of every mind, and other departments of learning were for a time crowded out." The class in botany helped to write a *Manual of Botany for the Northern States*, published in July, and the college honored its lecturer with a master's degree.

Eaton probably could have stayed at Williamstown, where everyone from president to freshman student was delighted with his work. But he had resolved to continue his own training—to learn "all the botany and mineralogy about this country"—and that could not be done if he settled down. In September, therefore, he left the college to cross and recross the "primitive" mountains, mapping the geology of a district fifteen miles in width. By walking and staying at farmhouses he was able to travel two hundred miles at a cost of three dollars. The tour ended at Northampton, where he lectured for seven weeks to a class of eighty people. It included four lawyers, three doctors, one judge, a senator, and a representative, as well as the governor's daughter. "My class-room," he wrote young Torrey, "is crowded with the first people here of both sexes and all ages."

From Northampton he went to other towns in Massachusetts and then to Albany. There he revised the *Manual of Botany* and lectured in the State House at Governor Clinton's request. Then came a course at Troy and another for legislators who came back to Albany in the winter of 1819. That course underscored the unconditional pardon, which by that time was three years old.

Eaton developed a fixed routine for organizing lecture "classes." His first step was to write some friendly local leader, who called upon other citizens. If several of these committed themselves, the next step was to advertise the course, its dates and its price, with a note telling where subscriptions could be left. Finally, Eaton himself appeared and, if necessary, gave an introductory talk such as the one advertised in the Pittsfield *Sun* for May 17, 1820:

BOTANY—A public FREE LECTURE on BOTANY, will be given by Mr. A Eaton, on Wednesday, the 17th instant, to commence at 8 o'clock, P.M. at the Lecture Room.

The object of this Lecture is to explain and to illustrate, by specimens, Mr. E.'s method of teaching practical Botany. Parents, Guardians, and all other Gentlemen and Ladies of this vicinity, are respectfully invited to attend.

This free lecture always attracted a few people who were unwilling to subscribe in advance at prices ranging from three to six dollars. In closing each course, Eaton secured testimonials which recommended him "to the Trustees and Instructors of Colleges and Public Schools," stressed his "indefatigable zeal and industry," or declared that his subjects (botany and geology) were "a very useful part of female education" and of value to "all Agricultural Societies, as well as to private gentlemen."

There was no doubt in Eaton's mind as to the value of this work, which he carried on until 1824. In the first place it brought him a living: more than twenty-one hundred dollars a year for something he liked to do. Second, it put him in touch with local enthusiasts like those two professors in Williamstown. Third, it spread an interest in science among people who were just beginning to have leisure for reading and serious

Ex-Convict Professor

thought. Nor did Eaton worry that such efforts called for showmanship.

"I do not know a person in the world but myself," he told Torrey, "who would become a successful scientific pedlar. I have learned to act in such a polymorphous character, that I am, to men of science a curiosity, to ladies a clever schoolmaster, to old women a wizzard, to blackguards and boys a shewman and to sage legislators a very knowing man."

In botany Eaton agreed with the ladies, but took his stand with the legislators when he spoke on geology. His first work in that field had been done in prison, where he copied more than three hundred pages from Hutton's unfair critic, Kirwan. Silliman inspired him to original work, and in 1818 he published a small textbook, an *Index to the Geology of the Northern States*. A year later he felt "vain" of his industry and success, and in 1820 declared—again to Torrey:

I have now ascertained, to my full satisfaction, that I am the only person in North America, capable of judging of rock strata. Silliman does not know how to distinguish the *old red sandstone* from the more recent (breccia), nor *puddingstone* from breccia, nor greywacke from greenstone trap. At least he has committed most horrible mistakes in all these cases. . . . McClure [Maclure] is much the best I know of. His book is a heterogeneous thing; but I can find a better application to facts in his book than in all american works. No person ought to write a syllable on geology, until he has seen the rock he speaks of in fifty or a hundred localities and compared all its various appearances and the various heads embraced in it.

Like most other geologists of those days, Eaton's ideas were biased by the doctrines of Abraham Werner. He thought granite and soapstone were precipitates; crammed all sandstone into a Secondary series, which also included gypsum, rock salt and ordinary limestone. Basalts and traps he grouped into one series, as the products of volcanic outbursts during relatively recent times.

Eaton had dwelt on religion while a convict, and he now took pains to make earth science agree with Genesis. The

The Story of the Great Geologists

planet began, he said, as a ball of pasty mud in which metals soon settled to form a solid core. Around this granite was deposited, followed by successive shells of Transition and Secondary strata. Intense heat developed beneath those shells, making steam that exploded and heaved granite upward till continents and islands appeared. On these dwelt antediluvian races which were destroyed when continental foundations gave way, producing the Deluge of Noah. For about a year earth's remnant of land-dwelling creatures floated in the Ark waiting for the flood to subside.

One reviewer was cold toward the early steps in this series, but warmed toward the later ones. Mud and metals were doubtful, the critic concluded, but when Eaton undertook to show that the "geogony of Moses and his account of the flood, do not in the least contradict the facts which experience has revealed, when he proves that the days of creation have been periods of time, as many learned divines have asserted, and every geogonist believes; we find him engaged in a desirable act of conciliation between science and religion."

The first edition of the *Index* was not a success, for barely eight hundred copies were sold in sixteen months. In 1820 Eaton brought out a new edition of 286 pages, priced at "1 dollar 50 cents." It contained a *Grammar*, "by the aid of which any person of ordinary acquirements" might become a geologist, explained that the theory of explosions was nothing but an aid to memory, and grouped fossils under nine Greek headings made by adding *lite*, or "stone," to their names. Thus fossil plants became "phytolites" while the awkward term "mammodolite" was applied to ancient mammals such as the mastodon. Eaton apparently knew little or nothing about William Smith's work and continued to group formations together if their rocks looked alike.

In this year, too, Eaton made a geologic and agricultural survey of Albany County under Stephen Van Rensselaer, New York's greatest landowner. It was followed by a study of Rensselaer County, a survey which ranged from "Transition"

Ex-Convict Professor

rocks to the care of apple trees and pigs. The "patroon" was so much pleased with it that he hired Eaton to survey a strip of country along the then new Erie Canal.

Eaton began work on November 11, 1822, traveling to Buffalo in a one-horse wagon with a driver and handy man. Field work was done between May and August of the next year, with time out for a series of lectures at the Troy Female Academy. By April 1824 the geologic part of his report was printed—163 pages with a plate of sections that stretched from Boston to Lake Erie. It was four and one half feet in length; merely to engrave it had "cost the Patroon 530 Dollars." The whole report sold for $1.50—a great bargain, said advertisements which placed the whole cost of the survey at twenty-five hundred dollars.

Eaton was cockily proud of his book, saying that it contained more geologic facts than all other American works put together. It did describe thick series of formations and made a badly needed start by giving some of them new names instead of borrowing from Europe. But the descriptions dealt with little except rocks; though fossils were mentioned Eaton did not use them to group or correlate related strata. Valleys were said to be made by solution, especially at times when the ocean overspread the land. Great masses of limestone and greywacke which seemed to be embedded in slate had fallen there, Eaton thought, while the slates were soft and pasty. As for briny springs: they were fed by supplies of liquid salt newly made by "nature's laboratory" somewhere within the earth.

The canal survey was followed by a colored map showing the economic geology of New York and parts of adjacent states. It appeared in a *Geological Text-Book* which Eaton published in 1830 and revised two years later. Here he recognized the Tertiary "series," described some fossils of value in correlation, and said that Lake Ontario's basin had been made by erosion. But he credited the Hudson Valley to "fusion by volcanic heat," and revived his theory of outbursts with improvements on its original form. These consisted mainly of two imaginary slices through the earth, one of which showed great stores of combustible substances under Primitive and later formations. In the next slice those stores had burst into flame, raising the Rockies, the Appalachians, the Alps, and other mountain systems.

Eaton also defended earth's records against churchly criticism, even while he used them as props for Genesis.

> Geologic facts [he declared], lead us to the history of created beings long anterior to written records. Such records may be erroneous, and we have no means of correcting them. But geological records are perpetual, unvarying, and can not be vitiated by interpolations or counterfeits. For example: The written history of the deluge might be varied more or less by erroneous copies and incorrect translations. But the geological records of divine wrath poured out upon the rebellious inhabitants of the earth at that awful period can never be effaced or changed. These later records add, to the Mosaic account, that even the antediluvial beasts of the forests and fens partook of the ferocious nature and giant strength of antediluvial man.

Eaton already had settled in Troy, where the Lyceum of Natural History had made him its official lecturer. He bought a large brick building called the Old Bank house and began to plan a school which would emphasize science and give its students practical experience as well as classroom instruction.

The first step toward such a school was taken in August of 1824 with a circular announcing courses for "operative chemists." The fee for each course was $25; "good plain board and lodging" could be had near the school for $1.50 per week. A "prudent young man"—or woman—need not spend more than $60 for the six weeks' term. "The above," Eaton's announcement continued, "is intended as a preparatory step towards a course of instruction in the general application of science to agriculture and the arts."

A second step was taken three weeks later when Eaton asked "Patroon" Van Rensselaer to advance three hundred dollars for apparatus and permit the school to be announced

Ex-Convict Professor

under his patronage. By November Van Rensselaer was paying bills, naming trustees, and appointing Eaton professor of chemistry and experimental philosophy (physics) as well as lecturer on geology, land surveying and the laws governing town officers. It appears, however, that the rich man merely signed papers after arrangements had been made by Eaton. On December 22, 1824, the latter could report:

> I have fitted up the 1st story of the Old Bank. I removed the partition and put in folding doors so as to open the directors long room into the 20 feet square room in the rear. Now the three great windows which look into the shop lock and the three great front windows, with our arrangement of cases, etc. give the whole the most elegant appearance of any school room I ever saw.

Thus began the Rensselaer School, later renamed Institute. Six students enrolled in January 1825 and began to follow Eaton's rule that the best way to learn was by doing. In chemistry they conducted experiments; in geology they worked with rocks; in surveying and engineering they made "plats" of farms, designed buildings and did other practical work. Since most of them would become teachers, none completed a course until he could lecture on and demonstrate its substance to the satisfaction of other students as well as that of Eaton.

Such methods provoke no surprise today, when "lab" work and practice teaching are found in every college. But they were both amazing and inspiring in 1825, when professors did little else than lecture from fully written manuscripts. Some critics pooh-poohed Eaton's "talking scholars," but others circulated petitions asking that new schools be established on the Rensselaerian plan. The first of these petitions appeared in Massachusetts barely six weeks after classes began at Troy.

Van Rensselaer's original loan grew faster than any snowball. Within three years he spent ten thousand dollars, adding ten more thousands before 1831. He bought equipment, paid tuition for county students, repaired buildings, yet never expressed impatience when new requests were made. He even helped Eaton out when a creditor demanded payments on the Old Bank place which the aging professor could not make. "As long as I am able and blessed," the patroon scrawled in haste, "you shall not be oppressed."

Still, the real burdens of the school were carried directly by Eaton. With little help except from students he taught chemistry, physics, geology, surveying and civil engineering, with courses in technology and other special subjects. He helped his sister-in-law instruct young women and guided the headmistress of a ladies' academy. He trained county schoolmasters, conducted tours, acted as the school's business agent, and supervised extension work. Between 1825 and 1841 he delivered more than four thousand lectures and made twelve thousand experiments in the chemical laboratory.

For all this, instead of being well paid, he sacrificed greater earnings and spent his own money. His lecture fees for one year had passed $2100; as professor he never received nor asked more than $1260, and after 1829 he was paid only students' fees plus tuition of special students, provided by Van Rensselaer. "My annual fees," Eaton wrote in 1831, "have never exceeded $800. I have supported my family with this and the profits I have made on textbooks. . . . I have been at full $3000 expense about the School, and about $700 of students fees are unpaid." In short, while the wealthy patroon was spending $20,000 the professor, a poor man to start with, sacrificed $7580 and paid out $3000 more.

Eaton did not complain, being one of those men to whom an ideal means more than a bank account. But he did have to ask for help, first from Van Rensselaer, then from his brother-in-law, and at last from a well-to-do son. When health failed from exposure and overwork, he also called in assistants who shared his devotion to teaching.

"Professor Eaton's health is very delicate," wrote one of these men in 1840, "he having frequent returns of bleeding at the lungs so that he is obliged to have an assistant of more experience than those which he has had for the last few years

Ex-Convict Professor

who came for their tuition and board, he offers me the situation and will probably give me considerable more wages than he has during the summer. Though if he did not I would stay."

The assistant did stay, and in January of 1842 announced a course of experimental lectures "in a central part of Troy." By that time the son, Daniel Eaton, had paid off the mortgage on the Old Bank property, the school had a "most excellent class," and Amos Eaton was planning meetings that would "give our citizens some knowledge of an institution ... which has furnished more than half the State Geologists of the Union, and a large proportion of the Civil Engineers and teachers of Natural Science."

A meeting was held on April 13; on May 10 Eaton died. One week later this notice appeared—a fitting memorial:

> Rensselaer Institute. This institution will *not* be suspended in consequence of the death of Professor Eaton. The Rev. Dr. Beman will exercise a supervisory care over its interests. Dr. Wright, one of our most distinguished naturalists, will assist in the Department of Natural History, and Professor Cook, the able and popular associate of Professor Eaton, will act as Principal. With such an array of talent, the Institute will prosper, we doubt not, as in times past, and continue to send forth to the world young men distinguished for their proficiency in the sciences.

CHAPTER XIII

Geologist at Large

WHEN EATON TRAVERSED the Erie Canal there was not an official geologist in the United States. Then, about 1830, came a change. Surveys were established in Massachusetts, Tennessee, Maryland and in several other states. The federal government sent an English scientist to the Missouri Valley, while the War Department encouraged Jean Nicollet to explore the upper Mississippi.

At last, in 1836, New York organized a state survey with a staff of four geologists and the same number of assistants. One of the latter was a youth named James Hall, who got his job only because Eaton insisted that he be employed. Within ten years he became the survey's chief survivor and was launched on a career that would make him one of the most influential, the most hated, and the most admired of American scientists.

James Hall was born at Hingham, Massachusetts, in September of 1811. His father was a woolen weaver and superintendent of a Hingham mill; his mother was an economical and able housewife. On her husband's small salary she maintained a family of five children in a house "on South Street in the heart of the village," and saw that enough money remained

Geologist at Large

for her husband to satisfy his ambition to become a pew owner in the meetinghouse.

Thus respectability was attained, even though economy demanded that young James be sent to the public grammar school rather than to the town's endowed institution. This seemingly was no loss, for the grammar master was an enthusiastic fellow who taught the boy zoology along with the three Rs, probably even helping him to dissect common animals of the seashore. For many years the two corresponded, the teacher acting as critic in scientific writing, but keeping peace by attacking the errors of colleagues, not of Hall. "We closet naturalists," he once exploded after a specially ill-written paper, "like to read the results of active observers in language we can understand."

By the time he was nineteen Hall had decided to avoid the respectable classics of his day and secure an education in science. The Rensselaer School seemed to offer just what he wanted so he borrowed money, walked to Troy and enrolled. He graduated in 1832 after doing excellent work in geology and chemistry. He at once set off afoot for the Helderberg Mountains to study strata and collect fossils; returning to Troy at the end of the trip, he awoke to the fact that he had neither money nor job, and that fossils, no matter how long their names, would not settle board bills. With characteristic haste he began to pack, ignorant of where he might go or what he might do. In the midst of the mess Eaton arrived, demanded an explanation, and hired him as librarian.

Thus, at twenty, James Hall secured his first academic post. He already was a B.N.S. (Bachelor of Natural Science); during the summer he became Master of Arts, and when the fall term began was adjunct professor of chemistry. The pay was small, and the work included such incidental jobs as tidying up the school and whitewashing one of its buildings. But young Hall did not mind these variations, and by 1835 was listed as a full professor.

During that year the law for a geological survey was passed

The Story of the Great Geologists

and signed by Governor Marcy, work beginning in 1836 with funds advanced by "Patroon" Van Rensselaer when New York's treasury ran low. The state was divided into four districts, one for each team of geologist and assistant, both of whom were to be young men. Hall was made assistant to Dr. Ebenezer Emmons in the Second District, composed of the Adirondack counties.

The plan was sound, yet things soon went badly. Geologist Conrad disliked strata of central New York, and vowed that outcrops might slide into Hades if only he were allowed to study fossils. Another district proved unnatural and unwieldy, while in the Adirondacks Assistant Hall was indifferently studying iron deposits and quarreling with his superior, Emmons. At the end of the first season districts were changed, Conrad was made paleontologist, and James Hall took his place as full geologist with a salary of fifteen hundred dollars.

Thus it was, in the spring of 1837, that James Hall set out to unravel the geology of western New York. He faced an enormous task: fifteen and a half big counties, largely covered by forest and with barely passable roads. On the other hand, formations were not much disturbed and their general sequence had been worked out during Eaton's survey of the Erie Canal. Wiseacres opined that the young geologist would pull through, but make few great discoveries.

They reckoned without two factors: the amazingly rich sequence of fossils and James Hall's independent mind. Petrifactions projected from beds of stone, were scattered among the pebbles of creek beds, and covered banks of weathered clay. Seneca Indians used hollowed cup corals for pipes, joined crinoid columnals in strings of beads, and treasured stony lamp shells as charms that were buried with fallen warriors. In a region where neither rock sequence nor structure offered great problems, Hall studied these ancient remains for themselves and as records of stratigraphic changes. They enabled him to recognize an almost complete series of Silurian and Devonian formations before either system was well established in

Geologist at Large

Britain. But instead of borrowing European terms he devised names of his own to designate groups of strata. America, he maintained, was such a long way from Europe that neither old marine basins nor sediments in them could have been continuous. When his report appeared in 1843 Hall even disregarded the terms "Cambrian," "Silurian" and "Devonian," in the belief that it was a "great misfortune that European systems were ever made a standard for our rocks."

This report, a quarto of 683 pages, was a notable work. On the practical side it gave usable descriptions of rocks, proved that coal was not to be found in New York, and so saved money for would-be miners and investors. In pure science it set up what for years was known as the New York System and gave the first clear, logical account of America's early formations. It established such native names as "Oriskany," "Niagara" and "Genesee" for series of related strata. It rejected floods and accepted glaciers, but followed Lyell's early opinion that floating icebergs had distributed great sheets of drift. It also followed Lyell by proclaiming that earth's history was one of gradual change, not the sequence of catastrophes pictured by Murchison and Von Buch.

The Survey ended when its reports were published, but Hall's half-studied fossils remained. Other collections had not been studied at all, for Conrad had given up in despair as four geologists and their aides sent in box after box of specimens. To make use of these a bill was passed authorizing a state "cabinet," or museum, as well as a paleontologist whose duty would be to study those fossils and describe them in an additional report.

Two men wanted the new position: Hall and Ebenezer Emmons, his one-time teacher and former superior on the Survey. Emmons was still drawing salary as a geologist, was a professor in the Medical College of Albany, and had influence with the governor. Hall possessed neither job nor money, and was trying to meet expenses by selling minerals. For a time it seemed that Emmons would win, but some chicanery with Hall

The Story of the Great Geologists

(whom he'd promised to help) and opposition from the Boston Society of Natural History told against him. When the rumpus ended Hall was paleontologist, a friendly collector had charge of the cabinet, and Emmons was tucked away as agriculturalist and professor of obstetrics.

Thus Hall triumphed, and at once applied those independent methods which on one hand enabled him to do his work, and on the other kept him forever fighting that it might be accomplished. Even before the bill authorizing a paleontologist was passed, he wrote that he had been "close to work for the last year," and had "between 300 and 400 species figured." Now, having pledged himself to complete the entire task in twelve months, he stopped writing and set out to make even greater collections. When the year was up he explained what had been done, asked for more money, and received it from the legislature.

At last, in 1847, Hall produced a report: not the one book on all the fossils of New York that had been promised, but a huge quarto volume devoted to those of the "Lower Strata" alone. It appeared, moreover, as Volume I in a series which, according to Hall's own plan, was to be the work of a lifetime. "I propose," he wrote a friend, "to cover the fossils of all the rocks below the Coal over the whole United States." Legislators grumbled at the project, of course, but Hall refused to drop it. Why should anyone ask him to abide by a promise which he had not intended to keep? And who were legislators, that they should block the path of science and James Hall?

So the work went on, and the first volume of the *Palaeontology of New-York* was followed in 1852 by a second, with a third already under way. Eventually a legislature, believing that all things should end, decreed that the work should be completed in eight volumes. Hall accepted the limitation—and then when it began to cramp, calmly divided volumes into parts and parts into sections, and went on publishing. In 1894, when Volume VIII was finally complete, the *Palaeontology* amounted to thirteen fat tomes, quarto.

Geologist at Large

Again and again the legislators balked. They cut down salaries and held up bills; sometimes they even stopped appropriations entirely. But less often than one might think, for the upstate politician who opposed Hall's schemes found himself in trouble. Not merely did he face quarrels in committee rooms and opposition on the floor; he laid himself open to assaults of a scientist who could use cutting sarcasm, attack with the fury of a circuit rider, and curse with the art of a river pilot. Rustic solons stuttered and blushed before his onslaughts while old-fashioned gentlemen and hard-boiled bosses took him to their hearts. Again and again they made the legislature give what he wanted, and year after year his wants increased. Eventually they passed eighty thousand dollars—and even then the money came.

Still, this does not mean that the state of New York paid for the *Palaeontology*. It didn't. Fully half of Hall's troubles came through his attempts to secure money and collections to be used in research for which the state got credit. While sniffers investigated, found that to describe certain species of fossils had cost an uncertain number of dollars, and therefore decided that the paleontologist was dishonest, Hall sold his lands and stocks to pay the wages of draftsmen. When the legislature held up appropriations and even refused him a salary, he barked and swore, but dug into his own pockets for money to hire assistants and buy collections that were needed for study.

But even this Hall could not do without raising endless rumpus. He had enormous needs for material, and his men had orders to get fossils, not make explanations. If they could not collect they begged or bought; what they could not buy Hall borrowed and did not readily return. Collectors complained and their friends protested, but Hall either could not or would not return useful specimens.

Thus matters went in Albany while the state paleontologist's fame extended through two continents. He was one of the first men whom Charles Lyell wanted to see in 1841. He was a founder of the Society of American Geologists and Naturalists and a member or director of several official surveys. Each new volume of the *Palaeontology* brought enthusiastic letters and reviews from both sides of the Atlantic. Hall accepted them without protests. Since he knew that his work was good, why feign modesty?

Acknowledged as leader, he became critic, and at least one critical venture took him into court. In November 1849 he received a letter from Agassiz, who had just seen "a monstrous map" published under the title of *Forster's Complete Geological Chart*.

> I do not know its author [Agassiz wrote], but I am so painfully struck with the crudeness of this production that I hasten to write you to ask if you will not join me in exposing publicly a work so full of false and antiquated views most childishly misrepresented, that its mere circulation would be considered abroad as a disgrace to American geologists. . . . This is the more necessary as it is a very showy sheet with tolerably well drawn figures, which might easily mislead ignorant men or directors of schools not conversant with the subject.

Forster was a village teacher, and his chart had been made for sale to schools. Hall found it in the office of the state superintendent of education, who, as Agassiz had feared, was impressed and inclined to give it his recommendation. The paleontologist borrowed the chart "for examination at leisure" and, with the help of Agassiz, condemned it so bitterly that the teacher-author brought suit.

The case was delayed, and while Hall and Agassiz prepared their defense Forster got the resentful Emmons to help him with a new and less erroneous edition. Printed quickly in Albany, it was shipped to New York by a Hudson River night boat. Hall heard of it, boarded the steamer and apparently threw the shipment overboard. At least, he reached New York next morning but Forster's charts did not.

At last the case against Agassiz was called. By a legal twist the whole point of the trial became a determination of Emmons' rank as a scientist, for Forster admitted that the super-

Geologist at Large

intendent of education would not have accepted his chart without a "certificate of recommendation of one Prof. Ebenezer Emmons, a man of approved knowledge, learning, judgment and skill in the art and science of geology." Agassiz promptly denied that Emmons was or had been such a man, and produced eminent scientists who upheld his contention.

For days twelve Irishmen and Dutchmen of Albany County listened to the geologic fireworks and doubted sleepily that it made much difference whether there existed a group of rocks to be known as the Taconic System. But they saw that only Emmons and the schoolteacher upheld it, while eight scientists and a mapmaker declared the Taconic a delusion and Emmons himself a fraud. When the talking stopped the jurors weighed this evidence and decided the nays had it. The case against Hall never was called.

Throughout the excitement, Hall pursued his normal, tempestuous career. Having chucked Forster's chart into the Hudson, he brought out one of his own which was a great financial success. He revived a long-dormant desire to publish a text, thus warning off upstarts; forgot it in plans for a great American University at Albany, with himself as professor of geology at fifteen hundred dollars for three months of teaching. The university failed and Hall turned to work for the Canadian Geological Survey as well as to speculation in land, mining property, lumber and vineyards.

All this was very necessary, for the legislature had stopped his salary though it authorized printing of his book. Hall went ahead with his own money.

For more than ten years [he wrote Agassiz in 1855], I have carried forward the department of Palaeontology with almost no aid from the State save the salary of $1500 which has been suspended since 1850 and I now have to depend on other sources of earning my living while I devote all spare time to the Palaeontology of New York. . . . During the ten years past I have expended for this object more than $20,000 beyond all that has been received from the State and now find myself reduced to the necessity of doing something to support my family.

The Story of the Great Geologists

That need was relieved in 1856, after influential men, "enlightened & educated" lawmakers, passed a bill to renew Hall's wages and put his work on a permanent footing. Meanwhile he had become geologist of the new state of Iowa, and with its governor planned both a survey and a state university. Hall selected assistants for the former and a president for the latter, and Governor Grimes went home well pleased.

Iowa, however, was not ready for such ventures into science and education. After three years of struggle college-grade work at the university was suspended; the survey existed with scanty funds secured by selling warrants on the state treasurer, while bankers levied discounts of twelve per cent on Hall's personal checks. When the first report appeared, with one volume on fossils and the other on the geology of counties that had little mineral wealth, appropriations stopped. Hall, however, refused to give up. By law he was state geologist and as such he functioned, sending out assistants when he could not leave Albany.

Iowans solved this problem quite simply. They let Hall keep his title and engage assistants, but refused to pay his bills. What if he did seize all collections of the Survey? Rocks and fossils meant little to people whose farms were being lost to land sharks and who could not buy wagons or shoes.

Until 1852, when the third volume of the *Palaeontology of New-York* appeared, Hall did his own scientific work while Mrs. Hall prepared his drawings. But as she gave up fossils for religion artists were employed, and when Hall assumed unofficial charge of invertebrate paleontology throughout North America the task grew much too great for one man. He therefore hired assistants and became a director of research.

These assistants inevitably were the source of trouble. During the fifties, sixties and seventies no man lived who could work in peace under James Hall. At once trustful, suspicious, jealous and domineering, he had the defects of both a dictator and an undisciplined child. Illness made him petulant; quarreling with artists, engravers and politicians, he worked off his fury on whoever was least able to escape or resist. He abused scientific competitors, punched the noses of his servants, and emphasized arguments in the laboratory with a shotgun kept near his table. Curses accentuated his opinions, while his trigger finger gave force to orders that this or that fact be written as it appeared to Hall.

Salary check from James Hall to Charles Schuchert, an assistant who later became a professor at Yale. The name and words "order" and "Fiftytwo" were written by Schuchert; the rest is in Hall's hand.

Under these attacks the first assistant had himself transferred to Missouri, where he felt it safe to resign. Others came and went, some to better paid jobs and others to any job that would take them out of Albany. Mere collectors fared better, for most of them lived and worked far away. Hall could only abuse them by letter, and hot words lacked the menace of fists or a loaded gun.

Amid these tempests Hall worked days, nights and Sundays, giving his young aides such training as no university could provide. His knowledge was theirs to draw upon; he paid them poorly yet gave them something to live on even when he himself had to work without salary. And if they left without quarrels of more than normal violence, he would do his best to get them jobs. Even those whom he attacked were not ruined, for scientists came to know Hall's failings and to discount his furies. To have been hired by him was an honor; if the en-

The Story of the Great Geologists

gagement ended in flight and denouncement, nature merely had taken its course.

But though Hall was generous with his knowledge, he was miserly with credit for work done under his direction. What assistants did was his own and his name, not theirs, appeared on the published report or paper. Now and then some specially

Proof that Hall wrote his own manuscript long after he engaged assistants. This first draft from Volume 5 of the Palaeontology of New-York *was written after 1885.*

able aide received junior credit for a study which he alone had written, while a few favored ones were mentioned pleasantly in prefaces. But in 1894, when the last part of Volume VIII of the *Palaeontology* appeared, its nominal author was James Hall, "assisted by John M. Clarke." Yet Clarke, who never belittled his superior, often said that he alone wrote the book, and that Hall did not see it until it was in proof.

These varied lapses gave rise to absurd stories—to tales that Hall never wrote a sentence, that he purloined state collections, that assistants who did his personal work were paid by public funds. Some of these are answered by official records; others are disproved by box on box of manuscript in Hall's

Geologist at Large

vigorous yet crabbed hand. In some of those boxes are bundles of checks; checks drawn on Hall's personal account and made to the order of draftsmen, secretaries and assistants who had a share in his work. They leave no doubt of financial honesty nor of the fact that Hall used his private funds for the public business of science.

Most of the reports that came from Hall's busy workshop were filled with descriptions of fossils. He wrote of snails, clams and sea scorpions; of jointed creatures related to the king crab and of plants so simple that some authors still refuse to accept them as organic. Yet geology proper was not forgotten, especially when Hall was employed by state and federal surveys. In 1851, for example, he described rocks of the Great Lakes region, linking some with strata in New York, but refusing to extend them to Europe. "The simplest principles . . . ," he insisted, "teach us that sedimentary beds, having the same thickness and the same lithographical characters, cannot have spread over an area so wide." Series might be, and were, of equivalent age, but formations were bound to differ in widely separate parts of the world.

Six years later Hall ventured into theories of mountain making. He already had proved that one sequence of strata extended from the Appalachian Mountains to the Rockies. Thin in the West, they thickened eastward till their measured total exceeded forty thousand feet. In the latter region they were crumpled and broken, forming valleys and elongate ranges from Alabama to Quebec.

It would be foolish, said Hall, to suppose that North America ever lay under forty thousand feet of salt water. Plainly the sea bottom had sunk, at the same time causing elevation in land not far away. The rising land thereupon was eroded, sending vast amounts of mud, sand and gravel into the near-by sea. There the sediment settled in strata which filled the sinking basin so rapidly that it could not become very deep. Depression thus raised adjacent lands, giving speed and power to their streams and assuring renewed supplies of sediment. But subsidence had two other effects: it stretched beds at the bottom of each basin till they cracked and allowed molten rock stuff to rise, forming veins or dikes. Higher strata, of course, were compressed and folded in a series of lengthwise wrinkles. Such folding, however, did not make mountains; they were carved by streams at a later time, after the whole basin had been raised by great continental movements. Nor did heat developed in uplift greatly modify rocks, since the prime cause of such change, or metamorphism, "must have existed within the material itself; [so] that the entire change was due to motion or fermentation and pressure . . . producing chemical change."

These ideas were not received kindly; one critic called them a theory of mountain making with the mountains left out. To this Hall reasonably replied that he had not tried to tell how this or that mountain range was pushed upward, but merely intended to show that mountain-making masses had been produced by sedimentation in basins. Somehow, then, those sediments were uplifted in continental masses from which rains or streams proceeded to carve ranges, ridges and peaks.

Today no one accepts Hall's idea that crumpling did not build mountains or that uplifts were always of continental scope. But the concept of sinking basins filled with sediment has become an essential part of modern geology. So essential, indeed, that we take it for granted and seldom or never credit it to Hall.

Passing decades brought influence and fame, but changed neither Hall's ways nor his temper. He took his meals in the house with Mrs. Hall but slept in the little bedroom of his "office," a building erected to house his collections when the state could provide no room. He still kept the shotgun above his table, sat on a piano stool while he worked, and dripped candle wax over books and specimens on many a sleepless night. Eventually, when the state provided space for paleontologic research in a public building, Hall changed quarters

Geologist at Large

under protest. The brick office at home fitted his needs; why spend a lot of time driving to Albany?

To legislators the aging, irascible man became a cherished tradition. They made fun of his topless gig, his decrepit horse, his stovepipe hat, and his breeches buttoned down the side in the style of 1830. They even said that with an injured leg and a rheumatic arm he was not dangerous, and they twitted him about his work. But when he thumped into a committee room and shook his cane under their noses, they gave him what he asked. Hall rarely thought the sums enough, since he asked for what he thought he could get, and never did they repay the money that he himself had spent. But it stands on record that in one year the state of New York appropriated eighty-three thousand dollars for the *Palaeontology*.

Opposition disturbed him less than it had done in earlier days. After half a century of independence, the white-haired paleontologist felt himself the concern not merely of politicians and scientists but of fate itself. When a senator haggled for figures on costs, Hall flattened him with learned testimonials; when an official vowed to have him discharged, he damned upstarts but did not worry. The next day brought news of the official's death, and Hall came to his office beaming. "Providence generally is on my side!" he called to an assistant.

Hall died in 1898, four years after the *Palaeontology* was completed and sixty-two after his first appointment as state geologist. He had founded American stratigraphy, had put invertebrate paleontology on its feet, and had kept it there when neither universities nor the federal government could do so. Youngsters whom he trained had grown into leaders, and his books gave firm foundation for work by leaders-to-be who were then in school.

He had, of course, no real successors. Entering science when even New York held bits of the frontier, he did not leave it until factories, suburbs and System had arrived. Even in the last decade of his life Efficiency had attacked and sought to have him ousted. He fought it bitterly, winning because of the power

The Story of the Great Geologists

he had built through a half century. But the men who followed him had no such standing, nor were they heirs to Hall's tradition of rugged individuality. They themselves belonged to the new order, in which scientists, like everyone else, must become standardized.

CHAPTER XIV

Immigrant Innovator

THE FIRST GOAL of New York's costly Survey was collection of facts which people of the state might use on their farms and in clay pits, quarries or mines. In substantial measure that goal was achieved as geologists mapped iron ores and salt deposits, proved the complete absence of coal, and outlined granitic outcrops with their zones of useful minerals. But such practical work was put aside when Hall settled down to write the *Palaeontology*. Ores, building stones, brick clay, profitable salt beds—why should he neglect his fossils for such workaday concerns?

Several state geologists followed his example, giving people who wanted to locate quarries or coal mines handsome volumes describing petrified fish teeth or calcitic plates of crinoids. No other course, these men felt, could be quite scientific in the sense of putting fundamentals first. How could one use systems or groups to trace valuable deposits unless the fossils that distinguished them were first classified and described?

How? Their question was answered by a young Scot named Owen as he rode back and forth through Indiana. Owen did collect and determine fossils, but took care not to let such work

[165]

claim the lion's share of his time. "I have considered it my duty," he once announced, "while surveying a country so new as ours ... to search out the hidden resources of the State, and open new fields of enterprise to her citizens. That object effected, time enough will remain to institute inquiries (which a liberal policy forces us to overlook) of a less productive and more abstract character." A sane as well as practical rule, and one which made the man who framed it America's first great economic geologist.

David Dale Owen was born near New Lanark, Scotland, in June of 1807. His father was the rich and famous Robert Owen, a reformer as well as part owner and manager of great spinning mills. In a day when most manufacturers paid as little as they could and drove employees from dawn until dark, Owen raised wages, shortened hours, established day schools for children, and led classes for adults at night. Though conservatives foretold ruin, profits began to rise and conflicts that plagued other mill towns left New Lanark untouched.

David Dale was Robert Owen's fourth son in a family of eight children. Their home was a big, rambling mansion surrounded by forty wooded acres which lay beside the river Clyde and were crossed by a deep ravine known—almost incredibly—as the Gullietoodelum. The estate offered playgrounds for every season while the river furnished swimming for both boys and girls. There was little play with youngsters of the town, but frequent association with reformers, statesmen and political theorists who came to investigate New Lanark. David Dale specially remembered a visit from the Grand Duke Nicholas, who held the nine-year-old boy on his knees and wanted to take him to Russia as protégé or adopted son.

Both parents refused the grand duke's request, for the trammels of royalty did not appeal to them. David Dale stayed in Scotland for eight more years until he and a younger brother, Richard, were sent to the famous Fellenberg School at Hofwyl, Switzerland.

Immigrant Innovator

Headmaster Fellenberg was a progressive educator, but one whose ideas differed greatly from those of progressive educators today. He undertook to "develope all the faculties, physical, intellectual and moral" with a program that got boys out of bed at five o'clock in the summer and six on winter mornings. Formal classes filled nine hours of the day; five meals took their share of time; music, gymnastics, dancing, riding, agriculture and mechanical training kept things humming till eight o'clock at night. Some variation in courses was permitted, chiefly during the third year. David Dale Owen—called Dale—chose to specialize in mathematics and natural science, both of which were thoroughly taught. Both he and Richard thought chemistry classes a failure, since the school lacked equipment for student experiments.

While the lads were in Switzerland their father's interests and their home underwent a radical change. A Quaker partner managed to stop singing, dancing and military drill in the schools of New Lanark; his interference greatly upset Robert Owen, who gave up management of the mill and began to look for a Utopia where faith could not halt good works. His goal seemed to appear in a town then called Harmony on the distant Wabash River in southwestern Indiana. It had been built by a religious group of Germans called Rappites, or Harmonites, whose leader now wished to sell and lead his communistic followers back to Pennsylvania. Owen crossed the Atlantic, bought the town and twenty thousand acres, and announced an ideal society in which all would labor, all would be learned, and no one need be poor. William Maclure undertook to provide scholars and a school system while Owen left a son in charge and dashed back to Scotland. There he moved his family to Glasgow pending a final shift to what now was *New* Harmony.

To Glasgow, then, came the Owen brothers after completing their work at Hofwyl in 1826. Enrolling at the Andersonian Institute, they learned that geology was a science and encountered a system of education that bore some resemblance to

The Story of the Great Geologists

Eaton's. Applied chemistry was emphasized in popular lectures as well as formal classes. The young men set up a home laboratory with the help of their sister, Jane, declaring that experiments taught them as much in a day as they had learned at Hofwyl in months. A year later, when they set out for Indiana, Dale Owen took boxes of equipment to be used in applied chemistry courses for the colonists.

But New Harmony in February 1828 offered little opportunity for popular education in science. There already had been dissensions which racked the colony's social structure and forced malcontents to leave. Yet people who dwelt in the Rappites' sturdy stone houses were an unstable mixture in which cranks, ne'er-do-wells and skinflints mingled with decent pioneers and savants whom Maclure had brought from Europe. While these latter groups labored the crackpots quarreled and the wastrels asked for more money from Robert Owen's purse. After spending forty thousand pounds he rebelled, divided New Harmony with Maclure, and returned to the British Isles.

While agents struggled to keep the town alive, Dale Owen rode through the countryside, worked in the printing office, and studied lithography. In 1830 he went to New York, added painting to lithography, and thought of sailing for Paris. A year later he and an elder brother were in England, helping their father with an institute for mechanics at which Dale and a young American, Henry Rogers, gave popular lectures on science while the former studied chemistry at the University of London. This lasted till June 1833, when Robert Owen wandered off on some new enthusiasm. The brothers and Jane returned to New Harmony.

Now came three years of hard and purposeful work. While the village itself slowly revived, Dale Owen remodeled a battered building into a museum, lecture hall and scientific laboratory. It boasted terrestrial and celestial globes, equipment for chemistry and physics, and "an adult skeleton, complete, with brass springs and joints." Owen used it in some of the forty

Immigrant Innovator

popular lectures which he gave during 1833-34 to audiences of about fifty townspeople. He also made extensive chemical experiments and analyses, carefully observed the weather, and began serious work in geology. He spent two terms of five months each at the Medical College of Ohio, in Cincinnati, studying physiology and anatomy for the help they might give in investigation of fossils. Although he took the degree of M.D., he did not intend to practice.

After one of his trips to Cincinnati Owen proposed to Caroline Neef, the daughter of a schoolmaster who had returned to New Harmony. She was eager to marry her graceful and talented young suitor, but replied with pre-Victorian primness that his offer was a great honor not to be accepted offhand. "Will you please," she concluded, "consult my respected parents, whom I obey in all things."

Owen had warned Caroline that he already was wedded to the remodeled laboratory. Still, it was field work instead of chemistry which cut short their honeymoon to Mammoth Cave and set hasty preparations afoot when they reached New Harmony. For after two years' delay the state legislature had ordered appointment of a "person of talents, integrity, and suitable scientific acquirements, as geologist for the state of Indiana." The salary was only fifteen hundred dollars, with two hundred and fifty dollars for expenses, but Owen was eager to do the job as a service to both the state and science. The appointment came eight days after his marriage, and he soon was in the field.

Owen was no reformer, like his father, but he did believe in applying science to the welfare of man. He saw settlers rush into what then was the West, building roads, cutting down forests and establishing town after town. He also saw that those towns were held back when too much money was sent to the East in payment for manufactured products that came from minerals. Westerners themselves should provide coal for smelters, iron for implements, clay for pottery, and stone for substantial buildings. When they did so money would stay at

The Story of the Great Geologists

home to pay workmen instead of buying things from the Atlantic seaboard or Europe. Money also would come from rich men in the East, who would seek investments as soon as the West could offer profitable resources. The geologist's job was to find those deposits, at the same time forestalling unsound projects whose failure would do great harm.

Such was the broad conception on which Owen based his work. He began by following the Ohio River, where he found that strata which seemed to be horizontal actually rose to the eastward but dipped lower and lower toward the southwest. This meant that Indiana's oldest rocks formed hillsides close to Ohio but vanished under younger beds as one traveled toward Illinois. In Perry County rocks of the Coal Age appeared —sandstones, limestones, shales and other deposits which were three full ages younger than beds on the state's eastern boundary.

Owen realized that Indiana's coal was not a separate deposit, but was part of a rich field that stretched from Kentucky across Illinois and into Iowa. He also made sure that coal beds never were found below a limestone crowded with screw-shaped fossils known as Archimedes. If a prospector found those "petrified screws" he need not dig below them in the hope of discovering coal.

This work was almost incidental, for mineral fuel was not nearly so vital in 1837 as it is today. Owen spent most of the spring and fall of 1837 running zigzag surveys in search of iron ores and good building stone. In December he rode six days through rain, mud and flooded streams to take his report to the legislature in Indianapolis. A compact document of thirty-four pages, it was written for non-scientist readers and began with an introduction to the principles of geology. It made such an impression that the House of Representatives asked Owen to give a group of lectures in their hall. More important, they defeated a Senate bill which would have stopped the state geologist's work to save $1750.

In 1838 Owen undertook to locate coal beds near iron deposits, as well as to find salt springs, clay and shale for brick-making, and durable sandstones and limestones. He went to Virginia for a clue to the salt wells, discovered more iron ore, and predicted that the region around Terre Haute would become the seat of major industries. Again this report made a great impression, but again there was opposition to continued

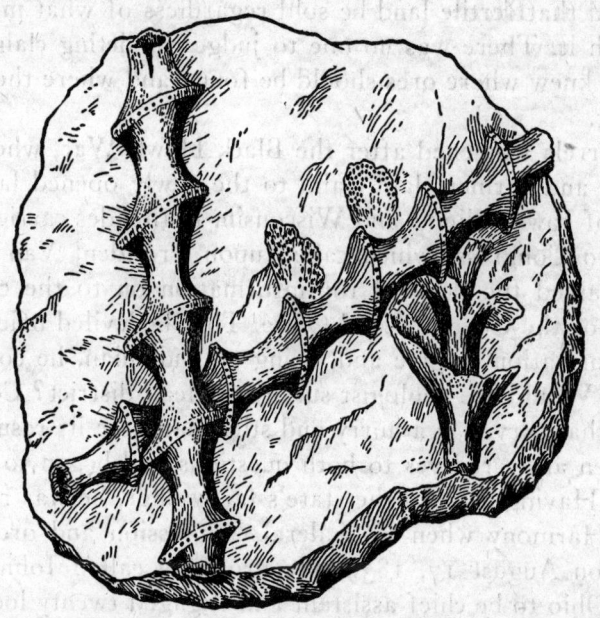

Limestone with specimens of *Archimedes*. Owen proved that coal could not be found below strata containing this fossil.

geologic work. When a bill finally passed the legislature it left salary and expenses unchanged but ordered the state geologist to examine soils, study silk and sugar manufacture, report on building materials for public works, and investigate milk sickness in cattle. Reappointment then was delayed until June, when Owen was so deep in a federal problem that he had to refuse.

Since 1807 Congress had followed a policy of leasing public mineral lands, though it sold those suited only to farming. The

The Story of the Great Geologists

scheme was a well-meant compromise, but one not equipped to work. Surveyors of the General Land Office laid out mining districts by guesswork or hearsay; ordnance officials wrote leases and gathered rents; rascals took oath that mining lands were agricultural or stole public ore between visits by collectors. Honest miners were in conflict with settlers who demanded that fertile land be sold regardless of what might lie beneath it. There was no one to judge conflicting claims, for no one knew where ores should be found and where they were lacking.

Quarrels worsened after the Black Hawk War, when both miners and farmers laid claim to the newly opened land districts of Iowa, Illinois and Wisconsin. Both sides carried their woes to Congress, which called upon President Van Buren, who passed the demand for information on to the commissioner of the General Land Office. That bedeviled official had no information to give and, being an Indianian, he consulted Owen. Would the geologist survey the lead district? Could he make that survey in a hurry and still guarantee its results?

Owen answered yes to both questions and began to lay his plans. Having refused the state's tardy offer, he was ready in New Harmony when his federal commission and orders arrived on August 17, 1839. He promptly called John Locke from Ohio to be chief assistant and engaged twenty local men who had learned some geology from his own popular lectures. With this skeleton staff Owen hastened to St. Louis, there to hire eighty more men and advance three thousand dollars for purchase of equipment, food and instruments. He taught his staff mapping and geology at night, as well as on the stern-wheel steamboat which took the expedition up the Mississippi. On September 17 the boat was tied to trees on the Iowa bank opposite the mouth of Rock River. One month after receipt of instruction, Owen's men were in the field.

Owen had never been a soldier, yet he carried out arrangements with military precision. Men and materials were divided; districts were assigned; tasks and times were settled. Soon

Immigrant Innovator

twenty-four parties rode out of base camp, each commanded by a subagent whose wage was four dollars per day. Unskilled assistants received two dollars, while Owen's pay was eight.

For this he, with the equally vigorous Locke, worked from dawn till dark and often after midnight. They met each party at stated times, checking upon the schedule of thirty quarter sections per day. They crossed the entire district eleven times, examined regions of special importance, and struggled against rains and deep mud. Several men fell so ill that they had to resign; Owen's death was announced in New York papers, giving Washington officials the jitters till his reports proved the rumor false. His worst disaster, he said, was staying up two nights to keep tents from blowing away. To prevent recurrence he hired a voyageur whose job was to pitch shelters so securely that only a tornado could blow them down.

Work meanwhile went forward at an amazing rate. Forty townships were surveyed in the first sixteen days, an average of ninety square miles per day. Work in Iowa was completed by October 24, and the Mineral Point region of Wisconsin required barely three weeks. Ten days sufficed in Illinois—ten days that came to an end on November 24. On the twenty-fifth "a severe snowstorm occurred, a gale blew up from the northwest, the thermometer fell to 12 or 14 degrees below zero, and the expedition could not have continued its operations in the field for a single day longer."

That storm marked the end of work for his men, but Owen still had months of labor in New Harmony. Until February he drafted maps; then he classified thousands of specimens and, finally, he wrote a report which linked pure and applied science. It declared that the lead ores would keep ten thousand miners busy, yielding more of this metal per year than did all the mines of Europe. It showed what soils were fertile and told why, proving that both miners and farmers could prosper side by side. It closed with a chapter on kinds and qualities of timber, written by one of the subagents who had led a field party.

[173]

The Story of the Great Geologists

Congress was prompt but niggardly; the report appeared in September 1840, without its illustrations or map. Dull-looking, it still caught the favor of a public dazzled and drawn by the West. Capitalists found it a guide to investment, engineers used it to plan mines, farmers welcomed it as proof of rich soil "out yonder" where land was plentiful and cheap. A guide for new settlers quoted from it, while an eloquent reviewer told how Owen beckoned readers to the "prairie wilderness" where elk herds stalked in majesty while the wolf sneaked through long grass "like a self-convicted culprit." Finally, fellow geologists voiced such a demand that in 1844 the book was published anew with sections, plates and geologic map.

Owen meanwhile was busy at New Harmony. He divided Maclure's collection, enlarged his own, and set up as a dealer in geologic specimens. He fitted out a larger laboratory, gave series of lectures, prepared a plan for the Smithsonian Institution, and finally located the stone of which it was built. In 1841 he took a trip down the Ohio, meeting James Hall at Louisville near a world-famous coral deposit. Hall slept on a box and so, doubtless, did Owen; the boat was crammed with specimens in boxes and barrels as well as with massive slabs. Owen's share exceeded two tons when the craft was unloaded at Madison, Indiana.

By 1847 Owen was ready to give up free-lance work for a position and salary. He applied for a Southern professorship, considered field work in Missouri, and urged a Kentucky survey. But the professorship had been filled and geologic bills bogged down in the two state legislatures. Prospects looked less and less hopeful till a Congress stung by new farm-mine conflicts ordered two expeditions sent out to what then was the Northwest. Owen chose the Chippewa Land District, which stretched from northeastern Iowa to Lake Superior. It thus lay just beyond the area surveyed in 1839.

The district was as large as New York; a vast wilderness of prairie and forest with only a few miles of road. The geologists used canoes, paddling up streams and along lake shores

JAMES HALL
as president of the American Association for the Advancement of Science, at the age of forty-five.

tesy of Noah T. Clarke

NEZER EMMONS
of Hall's rivals and
ter enemies.

DAVID DALE OWEN and a wood engraving showing a camp of his assistant, Evans, in the Badlands. Both from Owen's report of 1853.

Immigrant Innovator

amid "swarms of musquitoes, buffalo gnats, gadflies" and other voracious insects. Maps were poor and food was scarce, yet in five months Owen's staff finished their work and turned toward New Harmony. There the leader wrote another sprightly report which told of ores, described ancient limestones and sandstones, and showed how they overlay still more ancient granite. Sections, scenic plates and pictures of fossils illustrated the text.

Congress had planned one season's survey, but the General Land Office saw that more work was needed. One party therefore went back to Michigan while Owen spread his studies over the whole Northwest Territory during 1848–49. He zigzagged across what is now Minnesota and followed the Red River through country where the Sioux and Chippewas were at war. The mere mention of this route sent voyageurs scurrying, caused a pilot to stop at Ottertail Lake, and brought dire warnings of danger from peaceful Indians met on the way. With two inexperienced men Owen pushed ahead, shooting rocky rapids on the way to Pembina. There he learned that only one other white man was known to have made that trip.

To go back as they came would have been wasteful, so Owen and his men paddled downstream to the mouth of the Assiniboine. There they rested and bought supplies at the Hudson Bay post, were entertained by the governor, and secured an experienced guide. The party crossed upper Lake Winnipeg, took swift streams through the Lake of the Woods country to Fort William, and thence found its way back to Wisconsin. Before reaching Lake Superior Owen and his men made more than a hundred portages and shot rapids where "an error of half a canoe's length in striking a chute, or bringing to, below it, is sufficient to swamp the canoe, and expose to great peril the lives of all it contains."

Owen also planned to explore the Missouri but a rumored epidemic of cholera made boatmen refuse to go. Owen therefore sent an assistant out by steamboat and took his own party

The Story of the Great Geologists

to Iowa. While the delighted assistant collected fossil bones in the Badlands, Owen tramped across prairies, waded swamps and felt lucky to encounter bedrock two or three times a day. Yet he managed to collect fossils, trace series of strata, and

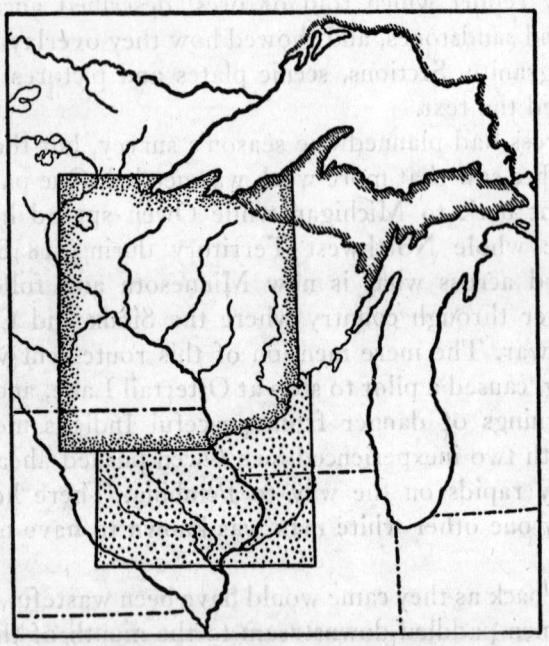

OWEN'S EXPLORATIONS IN THE NORTHWEST

Coarsely dotted area shows the parts of Iowa, Illinois and Wisconsin surveyed in 1839. Small dots mark the Chippewa Land District, surveyed in 1847.

follow promising coal beds as far as Fort Des Moines. There he sent his exhausted helpers home and struck out for the Missouri River, which he reached near Council Bluffs. His own work did not end until a leisurely steamboat tied up at St. Louis.

Owen labored three years on his report, a handsome quarto volume that was published early in 1853. It began with an account of field work, then described a thick series of formations

Immigrant Innovator

which ranged from Upper Cambrian ("Silurian") to Carboniferous. A specially interesting chapter recounted the discoveries of Subagent Evans in the Badlands and adjacent parts of South Dakota, then part of Nebraska Territory. There were reports by three assistant geologists, descriptions of fossil mammals and reptiles, and an appendix which dealt with fossil shells, crinoids, chemical data, plants and even birds. The report was illustrated by engravings on wood, stone and steel, as well as a "large Geological Map of the whole District, elaborately coloured" and engraved on copper.

Some congressmen opposed the report; others said that its cost—some $37,400 for nine thousand copies—made it one of the least expensive as well as the best of government documents. Leading journals of science reviewed it favorably and morocco-bound copies were sent to rulers of European countries. The one to Czar Nicholas of Russia carried a note that the author had been the little boy whom the then grand duke wanted to take home from Scotland back in 1814.

By this time Owen was forty-six; a slightly built man with a womanish mouth, a high forehead, and blue eyes deeply set. He dressed smartly in shades of gray, kept his Old World manners, and sometimes seemed foppish to people who had not seen him work. Those who had knew the power and drive which kept him going from early morning to midnight, inspiring his men and taking risks that halted devil-may-care frontiersmen. Sixteen hours were a normal working day; when stirred or hurried he worked eighteen to twenty and often forgot to eat.

Such days came often after 1855. Kentucky at last established its survey, with Owen as chief and two assistants who in time were increased to five. They mapped the state, sampled soils, collected fossils, and distinguished a series of formations ranging from the middle Ordovician to the Coal Age. The legislature approved such work so highly that appropriations were doubled in 1856 and increased again two years later.

A few people did grumble in 1857, when Owen stopped field

work in Kentucky and went to Arkansas. Two years later he shifted to Indiana, which appropriated five thousand dollars per year for a survey and asked that Owen direct it under any conditions that he might choose. He agreed to lay plans, manage accounts and supervise laboratory studies if his brother might carry on field work till that in Arkansas was completed. Three other assistants were added, and in the fall of 1860 Owen reported that he could, if desired, devote his whole time to this new and ambitious survey.

His desires, however, did not matter, nor did those of the legislature. Owen finally had worked too hard, bringing on a return attack of ague contracted in the Northwest. It was followed by malaria, rheumatism, a weakened heart and increasing digestive disorders. He was stricken again in October 1860, while dictating his second Arkansas report to three secretaries. Still able to think and speak, he kept on and completed the book three days before his death on November 13. "Poor Owen is dead, suicide!" was the way an assistant broke the news to James Hall back in Albany.

Suicide he undoubtedly was, but from success, not failure or despair. Owen had done the work he set out to do, increasing knowledge while he made earth science serve his fellow Americans. He had earned the friendship of farmers, miners, politicians; friendship which meant that geology was honored by those who paid its bills. Had others known how to do as well we should now be richer in good things of earth as well as in our knowledge of what to do with them.

CHAPTER XV

That a Nation Might Grow

WHILE OWEN PADDLED and tramped through the old Northwest another idealist pierced the wilds of eastern Canada. His goal, too, was useful knowledge—facts with which to develop a nation that did not yet exist. In this quest he discovered our continent's core and added unsuspected ages to earth history.

William E. Logan was born in Montreal on a chill April 20, 1798. Rumor has made his grandfather a Revolutionary loyalist who moved from Schenectady rather than give up allegiance to the British Crown. In sober fact, Grandfather Logan was a baxter (baker) who left the Scotch parish of Stirling to win fortune in Canada. Having money, he set up a large bakery in Montreal and bought land near the town. In the late 1790s he retired to a farm, leaving the bakery in charge of an elder son whose wife also had come from Scotland. Their third child was christened William Edmond, to honor both grandparents.

Little William began his education in the school taught by William Skakel, a good classical scholar who thought frequent whippings were essential to discipline. William paid close attention to his lessons and also learned to pass thrash-

The Story of the Great Geologists

ings on to older and larger boys. So self-reliant did the lad become that in 1814 he and a brother were sent off for further study at the high school in Edinburgh.

This was the school which Hutton had attended, as well as such men as Lord Brougham and Sir Walter Scott. For the time it was democratic; fees were not overly high and young noblemen recited with the sons of barristers and shopkeepers. The course was chiefly classical and lasted six years, but both Logan boys were so well prepared that they joined the fifth-year class. William soon led its two hundred students while his brother, Hart, was not far behind.

In 1815 the family returned to Scotland, settling in Edinburgh. A year later the boys graduated, and William spent one year at the university. There he studied logic, mathematics and chemistry but apparently was not inspired by either Playfair or Jameson. In 1817 he went to London, entering the business house of an uncle who also had come from Montreal.

For ten years young Logan remained in London, rising until he had full charge of his uncle's affairs. In spare time he studied drawing, read Homer and Cicero, translated French and Italian authors, and worked problems in mathematics. These he exchanged with a friend, each proposing problems to the other and solving those he received. This, wrote Logan, was a thoroughly "rational and useful means of keeping up an acquaintance." It left him time to attend concerts, play the flute, and on Sunday to hear "one of the worst preachers who ever wagged his head in a pulpit."

In 1831 Logan moved from London to Wales, where his uncle had invested ten thousand pounds in a process for smelting copper from discarded slag. The venture was not successful, nor was the smelter itself too effectively managed. Logan worked from six or seven in the morning till midnight, putting accounts in order and then managing the smelting of copper as well as the mining of coal. To do the last job well he studied the coal field and, quite without instruction, made a geologic map of it. Six years later he joined the Geological Society of

That a Nation Might Grow

London and exhibited maps of the South Wales coal district before the British Association for the Advancement of Science.

Logan's uncle died in 1838 and the nephew, now forty, resigned to study the origin of coal. By that time Werner's theory was forgotten, but a conflict raged between those who said that plant material had drifted into "pockets" and those who held that it settled more slowly and uniformly in prehistoric peat bogs. Logan confirmed the work of miners, who had found clay which looked like ancient soil under many coal seams. These underclays were filled with branching fossils called stigmaria, once thought to be strange water plants. Actually, said Logan, they were the roots of trees whose stumps often reached upward into beds of coal. All this evidence showed that plant material had settled in tree-grown bogs, where it partly decayed and then packed so firmly that it turned into coal.

In 1840 Logan sailed for America on a steamer greeted by marveling thousands as it docked at fog-bound Halifax. He studied ice and landslides near Montreal and in 1841 went to New York, traveling part of the way over a wooden railroad. He met Lyell on a street in the city, went to Philadelphia, and then studied underclays near Pottsville and Mauch Chunk. He was amazed at one coal seam fifty feet thick, worked like beds of rock in a quarry. He advised speculation in near-by coal lands, since farmers denied geologic evidence and gladly sold their fields for good agricultural prices. He predicted "incalculable" wealth for Pennsylvania, Ohio and the region which now is West Virginia because of their coal deposits and supplies of iron ore.

Back in Philadelphia, Logan talked with H. D. Rogers and other men who either were engaged in the geological survey of Pennsylvania or were stanch supporters of its work. Telling them how the British Survey had used his map of the Glamorganshire coal field, he aroused questions about Canada. It had, everyone knew, no geological survey, but Logan told how societies and committees had asked that one be founded.

The Story of the Great Geologists

The implication seemed obvious. Logan was a geologist; having returned to Canada, should he not conduct its survey?

This was in 1841, when Canada as we know it had not come into existence. The name meant only Quebec and Ontario, recently united in a queer, two-headed province whose capital for the moment was Kingston. In this combination Quebec was Lower Canada because of its position at the mouth of the St. Lawrence while Ontario, being farther inland, was Upper Canada. The provinces of New Brunswick, Nova Scotia and Prince Edward Island were distinct, fiercely jealous of their separate dependence on Great Britain, and determined not to unite with each other or their patched-up neighbor.

Logan ignored this schism by beginning study of the Pictou coal field, which lies in northern Nova Scotia. But it was the governor-general of Canada proper who appointed him director of the Geological Survey in 1842, after letters to recommend him had been written by Murchison, Sedgwick, Buckland and other great men of Britain. Logan was in England at the time but hurried across the Atlantic and up the St. Lawrence to Kingston. In the capital he was forced to dawdle while political storms died down. There was time for another trip to Europe before he could set to work in 1843.

The Survey appropriation was fifteen hundred pounds, for a time not specified. With it Logan hired an assistant and went to the Gaspé, that southeastern peninsula of Quebec in which coal had often been reported but where it had never been found. On the way he stopped at Joggins, Nova Scotia, where tides rise more than fifty feet in narrow Chignecto Bay and cliffs display tilted Carboniferous strata whose thickness is almost three miles. Logan measured the series bed by bed, finding seventy-six distinct coal seams and ninety underclays. Many of the latter contained thick roots of trees whose erect trunks reached upward through rusty sandstones above.

At Joggins one could walk at the foot of cliffs—when they were out of water. In the Gaspé Logan hired Indian canoemen, followed streams where they were passable, and where

That a Nation Might Grow

they weren't tramped over mountains or scrambled along rough shores. There he would stop at settlements of fishermen and buy lobsters for a penny each; inland he had to rely upon game and such foods as could be carried. He decided that the Gaspé sandstones were of Devonian, not Carboniferous, age and found supposed beds of coal on the coast to be nothing more than dark shale. He ended the season at Pictou, so ragged that he hid in a hotel room till the boat for Montreal arrived.

Logan's sketch showing one of his camps, with an open lean-to tent.

In the city Logan got another taste of political dallying. Parliament was not in session; no official was willing to provide quarters or otherwise aid the Survey. In desperation the director rented a house for laboratory and office, went to Albany for Hall's advice, and hired a young Pole as chemist to study minerals. Next year he took both chemist and assistant to the Gaspé, hired still more paddlers, and shot porpoises to save pennies spent for lobsters the year before. No one went hungry, but work was so hard that the Polish chemist soon gave out and went back to Montreal.

Winter brought its own uncertainties, greater than lack of quarters. The French party had lost power; its English opponents did not know their real strength in Parliament. Logan had spent his entire appropriation, had advanced the chemist two hundred pounds in wages, and had provided six hundred pounds for house rent and expenses. Parliament dallied while Logan talked to members and then drafted a bill giving the Survey eighteen hundred pounds per year. Friends rallied and raised the sum to two thousand pounds annually for a period of five years. The chief drawback to this bill—it soon became law—was that Logan must save enough from the increase to repay those eight hundred pounds of his own which were already spent.

These sums seem smaller than they were; a shilling went far in 1845, and Logan's salary of three hundred pounds was a good one when most people in Canada were poor. On the other hand, he had been paid more than thrice as much back in Wales for work not nearly so hard. Even while the Survey bill was being passed Logan refused an offer of twelve hundred pounds, "with the usual assistance," for work in India.

Why did he refuse? Partly, doubtless, because he was well-to-do; he could live without those twelve hundred pounds or any other salary. Partly because he loved the Canadian wilds as prospectors love Southwestern deserts or mountaineers love the Rockies. But chiefly, it seems, because he was deeply patriotic and felt that this work would help the people of Canada and might even guide a remote and blundering Parliament. Logan knew that copper had been discovered in Michigan's Keweenaw forests, the deposits trending toward land that would become part of Ontario. He had heard of rich iron ores, too, and of coal in the Saskatchewan country. By finding and testing such resources he might forestall errors like the one in which Britain abandoned its claim to lower Michigan after the War of 1812. Twelve thousand square miles of coal cast aside because of ignorance!

Logan declined the Indian offer in May; in June of 1845 he set out to explore the upper Ottawa River and Lake Timiskaming. Next year he went to Lake Superior, crossing to the Keweenaw Peninsula for a study of its copper ores. He examined one nugget said to weigh eleven tons, went into mine after mine, and followed veins made where waters had de-

That a Nation Might Grow

posited copper as they seeped from ancient lava flows. He then returned to the British shore of Lake Superior, where he interrupted geology to judge disputes between would-be miners and speculators in land.

Those trips of 1845-46 gave Logan his first close contact with rocks of great antiquity. He had seen some of them before, it was true; a thick series of pinkish-gray syenites in rounded remnants of mountains—the Laurentides—which lay northward and westward from Montreal. But at Lake Timiskaming the formations were more varied, beginning with belts of layered granitic rock, called gneiss, between winding bands of marble that seemed to stand on end. Above these came slates and conglomerates: ten to eighteen thousand feet of them in varied combinations. Lake Superior's shores showed true granite as well as pinkish, crystalline syenite which graded upward into gneiss. It was capped by a vast array of slates, greenstones, sandstones, lavas and solidified volcanic ash. Some of these beds contained copper, as did conglomerates which matched those of northernmost Michigan.

Logan could see that marble was contorted, changed limestone; he thought that banded gneiss and granite were remade sandstones and conglomerates. In his report on explorations of 1845-46 he therefore adopted Lyell's still novel term, "metamorphic," applying it to the oldest rocks on the shore of Lake Timiskaming. Seven years later he grouped these with formations of the Laurentide upland as the Laurentian Series. Rocks above them became the Huronian—perhaps, thought Logan, a New World equivalent of Sedgwick's Lower Cambrian.

These names carried implications and raised problems to which other men would give lifetimes of conscientious study. Logan willingly left the field to them, since his first goal remained that of useful service to the people of Canada. At one time he traced mineral veins near Lake Huron; at another he looked for iron ore or protected investors from the wiles of scalawags. Thus 1849 found him below Quebec, where promoters told of springs that carried chips of coal from rich beds underground. A little digging proved that the springs had been "packed," but Logan drew up a careful report to destroy promoters' arguments and convince the credulous. It showed that bedrock from which the springs bubbled had formed long ages before dead plants first made coal.

Once Logan did refuse a call, for an assistant whom he trusted had studied the region in which coal supposedly was found. Local people protested indignantly until the auger which brought up bits of coal also yielded bread and cheese. A workman busily packing the test hole had carelessly put in his lunch!

Such work Logan did for the public good, though the public did not always value it highly. One man sent ore samples with a shilling for analyses; if the latter pleased him he might try the Survey again. Another forwarded six pounds of rock said to be carbonate of lead and was furious when it proved to be worthless dolomite. The sample must have been changed en route, its owner argued, in a letter which then implied that Logan had been negligent.

> I did not send it [the sample] ... to have you *guess* what it contained. I sent it to you to have it assayed to know the percentage of lead and silver which it contained, in order to know how you agree with an assay ... made by other persons quite as expert as yourself.

By this time the Industrial Revolution was established and the scramble for world trade was on. Nations were eager to display their wares, while those rich in undeveloped resources made bids for immigration and investment. In 1851 Logan took Canadian rocks and minerals for display at an exhibition in London, receiving a salary but paying his own expenses, which amounted to four hundred and fifty pounds. Four years later he went to Paris accompanied by his mineralogist, Hunt, who had replaced the Pole. Day after day the two men worked from dawn until midnight, determined to make a good exhibit in spite of official indifference. They marveled at confusion and

That a Nation Might Grow

delays, at destruction of exhibits, and were amazed at "respectable" Frenchmen who either knew nothing at all of Canada or thought it was in Peru. Exhausted, Logan left France late in December 1855. In January he was knighted at Windsor Castle, the first Canadian to be so honored for work done in Canada.

Other European exhibitions came in 1862 and again in 1867. Survey work meanwhile went on, sometimes with increased funds and for some years with appropriations reduced to twenty-five hundred pounds. Since this paid salaries but not expenses of field work, Logan and his staff began to revise early reports as a committee had advised them to do in 1854. But revision alone would have been a makeshift; the real need was for a new book which would be unified and comprehensive as well as up to date. Logan began to write it with generous assistance from his mineralogist.

A fat volume of 983 pages, the *Geology of Canada* appeared in 1863. It began with a geographic account, less graphic than the descriptions of Owen, but clear and to the point. It then dealt with formations from Laurentian to Carboniferous, proposing new names for a few but using terms coined by Hall where they were suitable. Potsdam, Trenton, Utica, Medina, Hamilton, Chemung—these and other formations were shown to extend from New York into Canada without fundamental change.

Almost half the work was devoted to minerals, rocks and economic resources, which ranged from ores of silver and iron to strata suitable for grindstones. A final chapter dealt with many things: Laurentian and Ordovician strata, crumpled rocks near Lake Champlain, the Gaspé Peninsula. It emphasized records of the Ice Age, for Logan had joined Agassiz as a champion of glaciation. He described hillocks worn smooth by moving ice, rounded grooves that ran uphill, and scratches cut by sharp bits of stone as glaciers pushed them over bedrock. He concluded that the Great Lakes filled basins "not of geological structure, but denudation; and the grooves on the

The Story of the Great Geologists

surfaces of the rocks, which descend under their waters, appear to point to glacial action as one of the great causes which have produced these depressions." Final pages told of limestone boulders perched upon granite, of moraines that crossed valleys and dammed streams, and of clays that settled in lakes once fed by water from melting ice sheets.

SCENE ON THE RIVER ROUGE

Shows the rounded Laurentide Mountains, from which Logan named the Laurentian "System." From a notebook of 1858.

The *Geology of Canada* was compact, sturdy and attractively printed; Logan himself had advanced the money for new and attractive type. A colored map on which Hall had collaborated offered a much greater problem, since it seemed that work must be done in Paris under careful supervision. Logan devoted the spring of 1864 to a new Survey bill which both increased shrunken appropriations and guaranteed payment of his own advances, then more than seventeen thousand dollars. This accomplished, he reluctantly sailed for Europe—and in London was happy to find a printer who would undertake the map. All went well for a while, and Logan was able to attend meetings, take geologic trips and make visits before work began to bog down on problems of printing in color. It

That a Nation Might Grow

still lagged in 1868, when Logan hurried back to London and gave so much help that the long-awaited map appeared in 1869.

By this time the old Province of Canada had vanished, wrecked by dual leadership and rivalry between the English and the jealously conservative French. In its place had come a unified dominion, with five provinces and undivided lands stretching westward to the Pacific. Logan watched anxiously while statesmen planned and compromised, while conservatives grumbled and radicals hoped, and while a doctor turned politician overcame suspicious opposition in isolated Nova Scotia. The geologist must have felt relief, too, after Britain's Parliament ratified the new federation with less concern than it might have shown over some small change in English boroughs. Canadians might or might not chart their course well, but they surely would not dismiss it as routine detail!

Sir William had passed sixty-nine when Dominion Day first was celebrated on July 1, 1867. He had long planned for a geologic survey whose scope and staff would be equal to the needs and resources of a nation with the more than three million square miles of British North America. Much of 1867 had to be spent in work for the Paris Exposition—work so constant and detailed that it permanently injured his eyes. In 1868, however, Logan went to Pictou, carrying the Survey into a province still skeptical of federation and at the same time resuming work begun when he came back from Philadelphia in 1841. His energies and spirit soared, only to decline again in the tedium of Montreal. It was plain that the new dominion needed a younger and more buoyant head for its Geological Survey.

There were protests, but Logan's mind was made up. In January 1869 he resigned after recommending as his successor a man who had done good work in Australia. The appointment was not made for months, but with spring Logan went to England and Wales, crossing the ocean again in June to examine early rocks of Massachusetts. During 1871–73 he reviewed formations in Quebec, an unwelcome task but one that had to be done in answer to criticisms of his early work.

By this time Logan, like James Hall, had become almost a national legend. Miners, farmers and high officials knew him: a small, sprightly man who could not be discouraged; who drove himself unsparingly; who loyally supported good assistants but had small patience with dullards. He wore the long hair and untrimmed beard of a frontiersman, bought ill-fitting suits which seldom were pressed, and in cities wore the same high, sturdy boots that served well on field trips. When one pair of these boots gave out a storekeeper who was new in Montreal felt so sorry for the shabby old backwoodsman that he cut the price of new ones in half. Soon a well-known official hurried in for a pair of those five-dollar boots. *Five-dollar boots?* Certainly—a pair of those good ones which Sir William Logan had bought less than an hour ago!

Logan spent the winter of 1873–74 in the country near Montreal, where he read and often wept over sentimental novels. With summer he went again to Quebec's Eastern Townships, but sailed for England in August and spent the winter with a sister in Wales. At her home he became so gravely ill that a nurse was engaged and someone sat with him at night.

Spring brought a little improvement; though too weak to walk, Logan sat up, wrote, and enjoyed the garden from a wheel chair. But soon a specialist had to be called; a man of great reputation who could do no more than shake his head and praise the local physician. Logan humorously contrasted the great man's bill of £190 with the £57 13 s. 6d. for more than three months' service by the village doctor. There was just one way to square accounts, he decided: pay the smaller fee as it stood and then add a gift of a hundred pounds which the doctor need not divide with his partner. As thanks were stammered Logan's eyes misted. It was almost his last bit of joy before death closed his career in June 1875.

Sir J. William Dawson chief defender of *Eozoon* as a fossil.

er Merrill

[Tw]o specimens of *Eozoon*[;] in circle, an enlarged [secti]on of three stony layers [fr]om another specimen.

Sir William E. Logan and some of the crumpl[ed] marine rocks which [he] studied on the Canadi[an] Shield.

Courtesy Bureau of Geology Topography, Canada.

CHAPTER XVI

The Canadian Shield

WHEN LOGAN BEGAN to work in Canada geologists still talked and wrote of the Primary, or Primitive, System. In theory it contained earth's earliest rocks: those formed during a long problematic age that preceded the Cambrian. In practice the Primary was essentially where Werner had left it—a crude catchall for deposits that really were ancient and for others that merely appeared to be so. The result was a vast and orderless array of old granites and others not so old; of syenites, lavas, slates and quartzites; of marbles, gneisses and gleaming schists. New formations were added to it because they lacked fossils or because they were so much metamorphosed that their age seemed obvious.

Then Logan entered the Laurentides, paddled around Lake Timiskaming, and visited Lake Superior. In each region he found thick series of rocks which were older than the Cambrian but by no means such a chaotic jumble as the term "Primary" implied. Nor, for that matter, could many of them find their places in one beginning chapter of earth history.

We have seen how cautiously Logan proceeded; how at first he merely put what seemed to be the lowest, most ancient gneisses into Lyell's metamorphic class. They were, said he, profoundly metamorphosed sediments whose grains had been changed by heat and pressure deep down within the earth. Lyell had used the term "hypogene," meaning *deeply* or *nether-formed.*

Neither term told anything about age, as Lyell had emphasized and as Logan began to regret. In 1854 he admitted that metamorphosis was "applicable to any series of rocks in an altered condition, and might occasion confusion." To avoid this and follow the system of local names which had come into use on both sides of the Atlantic, he termed the gneisses and related rocks Laurentian. Though the newly defined series was widespread, it was based especially on rocks of the Laurentide "mountains."

Such was Logan's first attempt at order among Primary rocks. It seemed radical and needless to some critics but was welcomed by Hall, Murchison and other geologists. Next year Logan and Hunt together proposed the Huronian System for formations which were not greatly metamorphosed and which lay above the Laurentian at Lake Timiskaming as well as north of Lake Huron. The authors were not wholly confident of its age, however, for though their new system was "beneath the Silurian terrane" it had not yielded fossils. When these were found the strata might "well be referred to the Cambrian system (the lower Cambrian of Mr. Sedgwick.)" The thickness of the Huronian was given as at least four thousand meters, or more than twelve thousand feet.

In 1863 terms were reversed, for the *Geology of Canada* made the Laurentian a system and the Huronian a series. More important, Logan and his aides put the Huronian far below the Cambrian Potsdam Sandstone, which Hall had recognized and named in New York. Between these two lay the "Upper Rocks of Lake Superior," a vast series of lavas interbedded with red shales, sandstones and conglomerates whose deposits of metallic copper were to make Alexander Agassiz wealthy. Logan had seen these formations in 1846,

The Canadian Shield

and United States Government geologists had attempted to explain their conglomerates as fragments torn from earth's crust when lavas came rushing up from the deep interior. But the series went nameless for thirty years until a mining geologist called it the Keweenawan, from the peninsula of northernmost Michigan on which its rocks were best exposed.

Each of these advances was made with caution; each aroused criticism and had to be revised. In 1865 Logan himself divided his Laurentian by naming three formations of "limestone"—they were largely impure marble—which lay between bands of gneiss. His successor tried to make those limestones Huronian, along with lavas and sediments of the Keweenawan Series. That attempt was unsuccessful, but a really vital change came in 1885, when a young Canadian named Andrew Lawson announced that Laurentian gneisses and granites were not Canada's most ancient rocks.

What were they? Enormous intrusions, said Lawson; masses of hot, doughy rock called magma which were forced up from depths of the earth while mountains were crumpling at the surface. On their way the magmas filled folds or cut through sediments, as well as volcanic deposits that included both lava flows and bouldery agglomerates. None of the molten stuff reached the surface, however, but cooled and crystallized while thousands of feet underground. There some intrusions have doubtless remained, though others were brought to light as erosion destroyed the mountains which once covered them.

So much, then, for Logan's "basement complex"; it really was secondary. First honors in age belonged to those sediments and volcanics which were crumpled into mountain ranges while Laurentian magmas ascended. Although these formations were contorted, shattered and changed almost beyond recognition, Lawson was able to group them together as the Keewatin System. Later authors have divided the Keewatin, naming formations whose proper order within it still cannot be told.

The Story of the Great Geologists

Do these changes, these uncertainties, seem puzzling? They are much less so than the great continental core, or "shield," which Logan, Lawson and their successors have explored. Beginning near the St. Lawrence and the Great Lakes, it spreads northward around Hudson Bay to include two thirds of Greenland and lesser Arctic islands. Northwestward it extends almost to Alaska, while southern outliers reach into Minnesota, Wisconsin and Michigan, as well as into northern New York. Much of its surface is forested, with shrubs and a thick ground litter hiding all traces of bedrock. Vast areas are scrubby woodland, muskeg or lichen-grown tundra, while Greenland is covered with ice inland from the sea. Over most of the region settlements are few and roads are non-existent, so that canoes form the chief means of travel for those who must keep in touch with the ground.

Such, then, is the Canadian Shield: two million square miles of wilderness which forms the stable nucleus of North America. Yet that stability has been relative; part now lies beneath Hudson Bay, while formations record the uplifts and depressions of almost two billion years. Strata have been broken, tipped, contorted, absorbed; highlands have risen again and again; streams and the ice of repeated glacial epochs have reduced them to plains or low hills. Today most of the Shield presents a gently rolling surface marked by convex ridges, parallel streams, and networks of interconnected lakes. Exceptions are deep, precipitous valleys where erosion has uncovered canyons scoured and then filled ages ago. Lake Timiskaming occupies one of these canyons, where the walls stand farther apart than they do a few miles downstream. Saguenay steamers follow another whose precipitous cliffs and deep water remind tourists of Norwegian fiords.

Formations of the Shield vary greatly, partly because they differed from the first and partly because metamorphism has changed them in radically different ways. The result of such diversity is confusion, with local names and deposits that overlap, formations that seem alike but are not, groupings and ex-

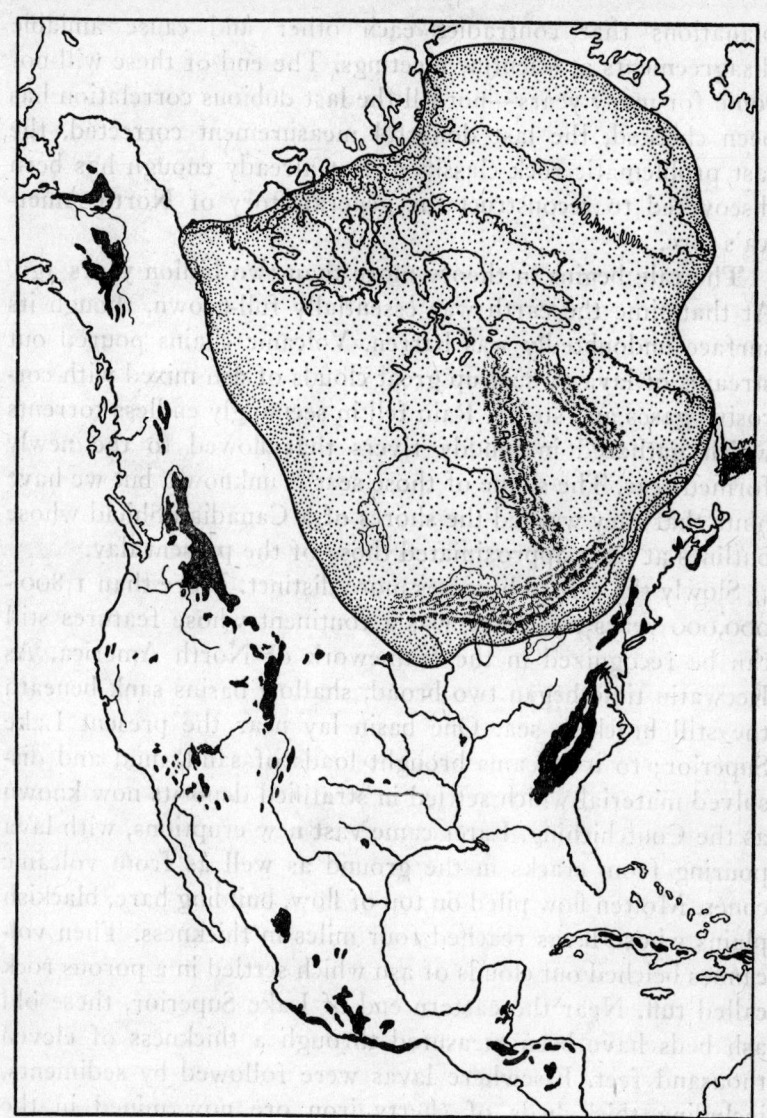

PRE-CAMBRIAN AND CANADIAN SHIELD

Pre-Cambrian outcrops in black; Canadian Shield stippled. On it appear three ranges of late Proterozoic mountains.

[195]

The Story of the Great Geologists

planations that contradict each other and cause amiable disagreements at scientific meetings. The end of these will not come for many years—not till the last dubious correlation has been checked, the last doubtful measurement corrected, the last problematic rock classified. But already enough has been discovered to reconstruct the general story of North America's core.

The tale begins in theory more than two billion years ago. At that time the earth was essentially full-grown, though its surface undoubtedly was barren. Volcanic chains poured out streams of lava or sent up great clouds of ash mixed with corrosive gases and steam. Rain fell in seemingly endless torrents which gathered in muddy rivers that flowed to the newly formed seas. The shape of those seas is unknown, but we have hints that they washed the shores of a Canadian Shield whose outlines at least approximated those of the present day.

Slowly the record becomes more distinct. More than 1,800,000,000 years ago there was a continent whose features still can be recognized in the framework of North America. As Keewatin time began two broad, shallow basins sank beneath the still brackish sea. One basin lay near the present Lake Superior; to it streams brought loads of sand, mud and dissolved material which settled in stratified deposits now known as the Coutchiching. Later came vast new eruptions, with lava pouring from cracks in the ground as well as from volcanic cones. Molten flow piled on top of flow, building bare, blackish plains whose lavas reached four miles in thickness. Then volcanoes belched out clouds of ash which settled in a porous rock called tuff. Near the eastern end of Lake Superior, these old ash beds have been measured through a thickness of eleven thousand feet. Elsewhere lavas were followed by sediments, including thick beds of cherty iron ore now mined in the Vermilion Range of northern Minnesota.

Eastward from the volcanic region lay a sea that covered parts of Quebec, Ontario and New York, probably reaching New Jersey as well. Its basin sank slowly through millions of

The Canadian Shield

years while sand, mud and limy ooze settled upon the bottom. Today these sediments are quartzites, schists and marbles which seem to be ninety-four thousand feet in thickness, but are so crumpled, broken and repeated by faulting that they may be a good deal less. Logan named them the Grenville, and in spite of some disagreement his name is in use today.

Deposition, eruptions, repeated deposition; these four words summarize events of the Keewatin Period. But at last the marine basins drained; new-made land warped into mountains, some of which crumpled and broke while others formed blisterlike ridges filled and interlayered with magma. Cooling, this rock dough crystallized into syenite and granite, while steam that worked its way through surrounding deposits rearranged or transformed their mineral grains, turning them into gneiss whose irregular bands recorded pressures in the crumpled ranges. These gneisses, with the syenites and granites, formed the complex series on which Logan based his Laurentian System. We now reduce it to the rank of series, but dignify the events which accompanied it as a revolution.

We limit the revolution to uplift and intrusion and follow it by one of those epochs known as "intervals of erosion." Yet such limits are arbitrary, for erosion began as soon as new land arose from the sea, to increase as frost shattered strata on rising mountains, as rain and snow pelted their surface, and as streams cascaded down their slopes. Rocks were broken to bits and carried away, apparently to seas whose basins still are undetermined. Destruction reached its height at the climax of revolution; it continued after folding and intrusion ceased, dividing the long, magma-filled uplands into ridges and peaks between which ran turbulent streams. In flood stage they swept sand and coarse gravel seaward, at the same time rolling massive boulders into valleys or onto piedmont plains.

How long this "interval" lasted no one can tell; millions of years may have elapsed while the mountains shrank into hills and foaming cataracts matured into winding rivers. Yet frost and rain still eroded the uplands, while muddy streams that swung to and fro reduced hills to an almost featureless lowland. Thus they completed what one explorer of the Shield has called the "most stupendous and widespread example of leveling known in the history of the world." Even critics who think that an epoch of still greater leveling was to come agree that millions of cubic miles of hard rock were removed from the continent's core during Post-Laurentian times.

A new period began as the epoch of erosion ended. At its outset the Shield was a low, level plain with here and there a hill that projected where rocks were unusually hard. A slight sinking admitted the sea, its waves mixing sand and mud with boulders worn from relict Laurentian cliffs. As waters deepened fine muds and lime ooze settled, to be interrupted by flows of lava and showers of volcanic ash. In some valleys streams piled up sand and coarse gravel; twenty thousand feet of such sediments settled in northwestern Ontario. The whole system is called the Timiskaming, from that canyon lake where Logan first saw it in 1845.

No one knows just what the Timiskaming contains, for its strata survive only as remnants scattered through a forested region a thousand miles or more in width. But there is no doubt that the period closed with another upheaval and interval of erosion. Again strata were folded, broken and compressed as parts of the Shield arched into mountains that equaled the modern Rockies in height. Again folded ranges were filled with pink and gray granite which Lawson named from cliffs that appear near the little town of Algoma. Those cliffs, of course, are mere roots of highlands destroyed during the epoch of erosion that came after uplift.

Mountain building, intrusion, erosion—these characterize the Algoman revolution, whose effects Lawson holds to exceed those of Laurentian changes. If we follow him, as many geologists do, we bring earth's second era—the Archeozoic—to an end with the destruction of Algoman mountains. If we wish to be more conservative, we may say that the Archeozoic ended with Laurentian upheavals and that the Timiskaming was the first age in the third, or Proterozoic Era.

The Canadian Shield

Such distinctions are not fundamental, for too much depends on what one means by the end or beginning of an era. What really counts is the fact that a basin north of Lake Huron sank and so initiated the Huronian Period. At first the basin contained swamps and monotonous flood plains which received sediment from streams that flowed out of what now is the United States. As time passed these low regions sank under salt water, only to rise again into land over which spread the glaciers of North America's first ice age. Melting after thousands of years, they left boulders as well as pebbles, sand and clay, all mingled in typical moraines. Such deposits, now hard conglomerate, range from one to six hundred feet in thickness and dot a region a thousand miles wide. Though less extensive than moraines of the ice age discovered by Agassiz, they still indicate glaciers much larger than the icecap of Greenland is today.

Scarcely had the ice disappeared when a sea that resembled Hudson Bay spread over much of the Shield. Its waves first piled up conglomerates; then, as waters quieted, great iron-bearing deposits were laid down. They were followed by thick formations of black mud which now are slaty shale. Their dark color comes from carbon—enough to make a hard-coal "seam" two hundred feet thick were it gathered in carbonaceous strata instead of being scattered through shale.

The Huronian Age closed rather quietly, with widespread uplift that raised marine beds but did not arch them into mountains. During Keweenawan times there was sinking while cracks opened in the ground and molten lava erupted. These eruptions built a basaltic plateau containing twenty-four thousand cubic miles of lava in the Lake Superior region, though even more molten rock spread out in subterranean sheets called sills. During intervals between eruptions, streams and lakes deposited fifteen thousand feet of conglomerates, sandstones and shales. Maroon to terra cotta and rusty in hue, these sediments form one of the greatest series of red beds to be found anywhere in the world.

The Story of the Great Geologists

Red beds have theoretical importance, since they are supposed to indicate arid climate in regions from which their sediment came. Of greater practical concern, however, is the fact that coarse red strata of the Keweenaw System are the ones which received metallic copper from cooling, solidifying sills. Additional supplies of copper, as well as silver and most of the world's nickel, are found in sills themselves and in old surface flows.

A final chapter in the Shield's Pre-Cambrian story began as three separate regions crumpled into mountain ranges. Molten granite oozed into the folds and, just as in earlier times, uplift was followed by erosion. Though not one of earth's greatest epochs of destruction, it still did away with the three mountain systems, one of which was at least seven hundred miles long and a hundred in width. Erosion also produced those deep canyons near the edge of the Shield, some of which are still partly filled with deposits of much later times.

As this record of early ages was deciphered, geologists had to abandon long-cherished misconceptions. First to go was the Primary itself: that theoretical age without divisions whose rocks could be lumped together in a single system. In its place appeared four great ages, or periods, divided among two still greater eras. Their importance in earth history may be gauged by figures given in the chart of Pre-Cambrian time—figures largely computed from radioactive materials found in unweathered rocks. They show that the Archeozoic began more than 2,200,000,000 years ago and lasted some 1,150,000,000 years, or about as long as all eras and ages since it came to an end. The Proterozoic, though shorter, stretched through more than 500,000,000 years, which is almost the length of time that has passed since the Cambrian began. The Keweenawan alone, with its sediments and lavas, represents 250,000,000 years.

Though the reality of early ages soon became plain, these figures which measure them in years were determined quite recently. Field work meanwhile disposed of another miscon-

PRE-CAMBRIAN TIMES AND ROCKS
IN THE CANADIAN SHIELD

	Divisions of Time and Their Length	Important Rocks and Events	Dates
PROTEROZOIC ERA		*Killarney, or Penokean Revolution*	
	Keweenawan Period 250,000,000 yrs.	Lavas, intrusive sheets and sediments. Great copper and nickel deposits.	565,000,000 620,000,000
	Huronian Period 250,000,000 yrs.	Animikie marine sediments, including shales, conglomerates and iron ores. Ash beds and hardened lavas.	
		Cobalt sediments, mostly non-marine; solidified glacial deposits.	905,000,000
		Bruce flood plain and marine sediments; some iron-bearing rocks.	1,035,000,000
ARCHEOZOIC ERA		*Algoman Revolution*	
	Timiskaming Period 300,000,000 yrs.	Coarse river deposits and shales with many different names.	1,050,000,000 1,090,000,000 1,200,000,000
		Laurentian Revolution	
	Keewatin Period 800,000,000 yrs.	Soudan marine sediments, including iron ores.	1,420,000,000
		Keewatin lava flows and ash beds with some sediments.	1,800,000,000
		Coutchiching marine shales, sandstones and impure limestones, now greatly metamorphosed.	2,200,000,000

FORMATIVE ERAS. Largely hypothetical; perhaps a billion years in length. **No divisions can be recognized.**

ception, which held that hardened Laurentian magmas were relics of earth's first crust. Until 1900 most geologists thought the earth had once been a molten ball, which began to solidify when its surface cooled and became hard. Since Laurentian rocks seemed to lie beneath all others, they were assumed to be relics of the original crust. Lawson corrected that error in 1885, since an original crust could not be intrusive in strata already formed. Later workers cast doubt on the whole notion of a crust by proving that both lavas and intrusive magmas came from pockets some miles underground rather than from the core of a still unsolidified earth. Today no one can be sure whether the earth was molten or not, but even those who defend a once-fluid globe make no claim that any existing formation is part of its first crust.

Most persistent of all was a belief which drew strength both from Wernerian doctrine and from the supposed molten globe. Werner's Primary was a lifeless age; not until Transitional times did earth pass "from its chaotic to its habitable state." A molten planet was even more surely lifeless, as was one whose crust had newly solidified. Nor did this condition quickly end, for the molten interior must cool by loss of heat that was conducted outward and then radiated from the crust. Theorists described a hot and unstable primitive earth on which falling rain turned into steam while molten rock repeatedly burst through the fragile surface. Even Cambrian seaweeds, lamp shells and crustaceans were supposed to have struggled along in tepid seas under constant threat of being parboiled or engulfed in fluid lava. So convincing were these word pictures that belief in lifeless Pre-Cambrian seas survived long after primitive crust and molten earth core had passed into limbo.

Logan took no part in these speculations, but he did attack the assumption that all very early seas were lifeless. In doing so he raised a problem over which learned men argued at length and with heat, yet which still lacks convincing solution after more than eighty years.

The matter began quietly enough in the early 1850s, when

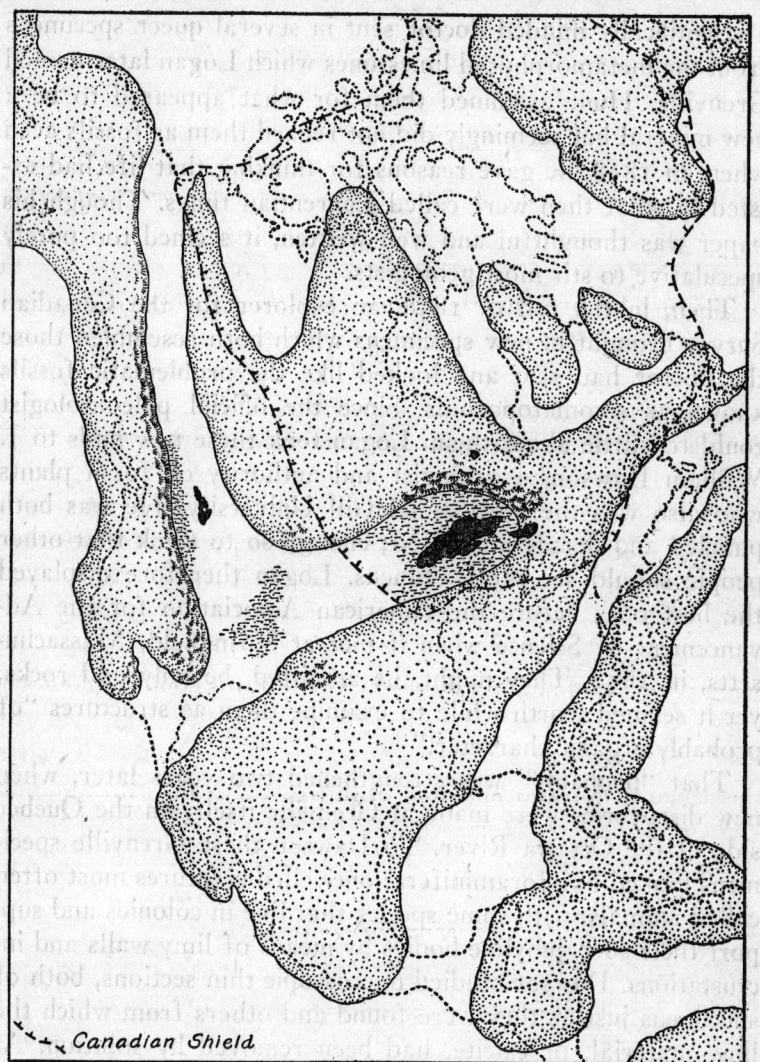

NORTH AMERICA IN LATE PROTEROZOIC TIMES

Stippled areas land; white areas sea or low land; black areas those of lava flows. Lined and dotted areas were land or very shallow water—perhaps both at different times. Note the outline of the Canadian Shield. Based partly on the work of H. C. Cooke.

a scientifically minded doctor sent in several queer specimens from the metamorphosed limestones which Logan later named Grenville. Hunt examined them for what appeared to be a new mineral but seemingly did not regard them as fossils even when, in 1858, he gave reasons for thinking that life had existed in what then were called Laurentian times. Though his paper was thoughtful and well written, it seemed too purely speculative to stir most geologists.

Then, in the fall of 1858, an explorer for the Canadian Survey brought in new specimens which both resembled those the doctor had sent and looked like the problematic fossils known as stromatoporoids. Since the official paleontologist could tell little about them, Logan took these new finds to J. William Dawson, a geologist and authority on fossil plants who also was the head of McGill University. He was both puzzled and deeply interested; enough so to think that other people should see the specimens. Logan therefore displayed the best ones before the American Association for the Advancement of Science when it met at Springfield, Massachusetts, in 1859. They might, he admitted, be only odd rocks, yet it seemed worth while to examine them as structures "of probably organic character."

That "probably" was strengthened two years later, when new discoveries were made at Grenville itself, on the Quebec side of the Ottawa River. To Dawson these Grenville specimens looked like foraminifers: one-celled creatures most often called animals, with some species that live in colonies and support their soft, jellylike bodies by means of limy walls and incrustations. Dawson studied microscopic thin sections, both of specimens just as they were found and others from which the limy material, or calcite, had been removed by solution. At last he decided to call the supposed fossils *Eozoon canadense*, or "dawn animal of Canada." The name was proposed in a group of papers which Logan read during his trip to England in 1864. Carpenter—the great W. B. Carpenter, who had studied "forams" for thirty years—added notes about structures in some newly found specimens.

The Canadian Shield

This combined announcement raised a stir which its authors hardly could have foreseen. Some scientists were enthusiastic; thus Lyell accepted *Eozoon* as the "greatest geological discovery" of his time, while a German thought it marked a new era in the study of fossils. Other critics were cautiously skeptical and a few became openly hostile. In 1866 two Irish professors declared that every important feature of *Eozoon* was purely crystalline. The ancient "fossil" was not organic at all, but a mixture of two minerals developed when impure limestone was metamorphosed into marble.

This attack drew fire from Carpenter, who announced new finds of *Eozoon* in slightly changed deposits of Hastings County, Ontario. He also remarked that one of the Irish professors knew too little about foraminifers to be sure what features were significant among their colonial forms. More important, the professor had cut no sections thin enough to show the very microscopic structures whose meaning he denied. Logan and Dawson came to Carpenter's aid with a detailed description of *Eozoon* from the almost unmodified limestone. Surely its structures were not products of ancient metamorphism!

Thus discussion turned into conflict, with Dawson taking the lead. A mild-looking, patriarchal man, his greatest interest had been education and repair of the breach between science and theology. A busy teacher, he still wrote abundantly—on ancient plants, geology of New Brunswick and Nova Scotia, Carboniferous vertebrates, prehistoric men, the last Ice Age, and popular geology. Thoroughly aroused by *Eozoon,* he studied hundreds of rock sections as well as others cut from modern foraminifers. He described, interpreted, explained; when these efforts failed to convince hostile critics he issued an invitation. Would they not come to Canada, where large collections could be re-examined with care, and where evidence could be seen in the field as well as under microscopes? For, after all, *Eozoon* was found in rocks, where its shapes, distribution and associations had real significance.

The Story of the Great Geologists

A reasonable suggestion, it seems, but one which most hostile critics rejected. The Irish professors stayed at home where they could look at their slides, repeat arguments, and announce with more aplomb than tact that Dawson might "employ his time much more usefully on other subjects than *Eozoön*." Another opponent matched details of Grenville specimens with structures in blocks of heated, altered limestone tossed from the crater of Monte Somma. Several American mineralogists made detailed microscopic studies, with results which showed that minerals in the original *Eozoon* were of metamorphic origin. From this they concluded that the entire "fossil" must be inorganic.

Still, Dawson held his ground to the last and in 1895 gave a series of lectures which showed that the problem should be studied out of doors as well as in the laboratory. He told how *Eozoon* always lay in domes, mounds or elongate reefs, illustrated its broadly conical shape, and described places where it had been broken into fragments which then settled in beds of pebbly limestone. In these fragments the "foraminiferal" structure was worn—ample proof that it had not been produced by metamorphism long after deposition.

Most significant, perhaps, was Dawson's comparison of *Eozoon* with columnar fossils found near Saint John, New Brunswick, and named *Archaeozoon*. For, though its name means "ancient animal," there now is small doubt that *Archaeozoon* belongs with things which we usually call plants —a limy, colonial structure built by several kinds of red or blue-green algae that lived together in one thick, slime-coated mass. While they grew lime settled upon the slime, forming bowl-shaped layers which soon hardened, preserving each colony in stone.

Such fossils are abundant in Cambrian and younger strata from Newfoundland to Virginia, among the northern Rocky Mountains, and southward to desert ranges of California as well as the prairies of Texas. From Montana to Arizona, they form beds and reefs in formations which seem to match the

The Canadian Shield

older Keweenawan, and so cannot be much less than 620,000,000 years old. Algal banks also are known in many parts of the Huronian, but some in northwestern Canada and southern Wyoming are more ancient, with ages reliably set at 1,200,000,000 and 1,450,000,000 years. These figures seem to make the fossils middle to early Timiskaming and very late Keewatin. Somewhat older, perhaps, are limeless algal filaments and cells found in cherty iron ores of northern Minnesota.

Pre-Cambrian fossils, therefore, are well known; those of Keweenaw age in the northern Rockies include shells, worm burrows and other traces as well as calcareous algae and a few seaweeds. Yet these things do not replace *Eozoon*. Is it really a fossil, earth's oldest foraminifer?

Most mineralogists still insist that the first finds of *Eozoon canadense* are metamorphic lumps in marble; lumps of minerals and nothing more. A growing number of paleontologists agree that the slightly metamorphosed specimens are algal colonies not much different from *Archaeozoon* and other well-established types. A few radicals—or are they conservatives?—suspect that the metamorphosed masses once were fossils, too, but have been changed so greatly that their nature appears only when banks or reefs are examined in the original rock. All camps are agreed on just one point: that *Eozoon* is not a foraminifer, as Dawson tried hard to prove.

There, in fact, lies the core of the problem. Dawson and Carpenter tried to prove that *Eozoon* was a one-celled animal; their opponents strove just as earnestly to show that it was not. In the conflict both sides sometimes forgot that their task was neither to prove or disprove a claim, but to ascertain all available facts and weigh all probabilities. Because they did forget we still await an unimpassioned, all-around description of what are either earth's oldest known fossils or minerals formed when a revolution remade impure limestones which are parts of the Canadian Shield.

CHAPTER XVII

Earth's Changing Time Scale

Eaton's *Geological Text-Book,* first published while Logan was still in London, divided all "regular deposites" among five classes which matched major divisions in time. From the topmost and youngest downward these were:

V. Tertiary. Strata containing fossil remains of "viviparous vertebral animals," or mammals. Most deposits were not yet solidified; few if any were hard stone.

IV. Upper Secondary. Strata containing fossil amphibians, reptiles and birds, all egg-laying vertebrates. Most rocks of this and older classes were hardened, or indurated.

III. Lower Secondary. Strata without vertebrate remains; fossils include snails ("univalves not chambered") as well as other invertebrates and relatively simple plants.

II. Transition. Strata in which the fossil animals are corals and other radially organized types, as well as bivalves and mollusks with chambered shells, related to the nautilus. Plants include seaweeds, both real and supposed.

I. Primitive. Rocks containing no fossils, and never found above fossil-bearing strata.

Besides this series, Eaton recognized "anomalous deposites" formed at the earth's surface by fusion or erosion of pre-existent rocks. Chief of these were volcanic deposits and diluvion, the former consisting largely of dark material that had hardened before Noah's Deluge, though recent lavas were included. The diluvion was glacial drift, supplemented by old river deposits and sediments that settled in lakes. Eaton attributed the whole class to deposition by water "in a violent state of action"—in other words, to Noah's Flood. Much younger were post- and annalluvions, both surface accumulations that must have followed the Deluge.

This grouping was far from adequate, yet it was better than Werner's and no worse than others in vogue before Lyell's *Principles* appeared. That work brought a rapid advance, as did papers by Sedgwick, Murchison and continental geologists. In spite of Hall's disapproval, European improvements were brought to America and adapted to American strata. A textbook popular in 1864 used such terms as "Silurian," "Devonian" and "Pleistocene," grouping related periods into four great eras. Of these the Azoic ("without life") was much like Eaton's Primary, though a map clearly showed outlines of the Canadian Shield. The author took pains to say that the name was not used literally, since he himself knew three reasons to suppose that simple organisms existed during Azoic times. So cogent did his arguments seem that he followed them with others to prove that the earliest living things were plants, not animals. "From this point," he concluded, "the progress of the life of the globe is a prominent part of geological history."

New articles and books continued this process of dividing and grouping both formations and the ages during which they were made. Thus the Silurian of 1864 became Cambrian and Ordovician, the polyglot Azoic fell apart, and Eaton's Upper Secondary became the Mesozoic Era, with three or even four periods. Parts of the Carboniferous were split off, and for a while the Tertiary was divided, with Lyell's epochs masquerading as full-length periods. In 1911 one man broke the Cam-

The Story of the Great Geologists

brian into three periods and inserted still another between these and the Ordovician.

Such refinements, however, went too far—and discussion of them may make the same error. Instead of tracing change after change, let us merely say that for a century after 1833 historical geologists were busy revising the scale of earth's ages while specialists in fossils and geologic structures sought new means of apportioning rocks. Their results were just becoming stabilized when physics and chemistry entered the picture with means for setting an age in years for this or that deposit. Thus a time scale which had been relative took on elements of precision.

Elements, not a complete aspect; for it is not easy to find unweathered rocks of known relative age which contain helium, uranium or lead of radioactive origin. Yet enough have been analyzed to give us a calendar of the past which sets general limits on eras and ages and provides some significant dates. These, in turn, may be correlated with important geologic events and with changes in earth's inhabitants. This calendar completes the one begun in Chapter XVI, which summarizes the history and deposits of Pre-Cambrian times.

Will there be further changes? Undoubtedly. Already we find hints that Pre-Cambrian periods must be modified, with rearrangement of formations. Laboratory workers are computing new and increasingly reliable dates, while field parties find better and better samples or determine with greater accuracy the limits of periods, formations and rock systems. Special attention is often given to revolutions, whose length may have been much greater than today's data indicate.

Still, these improvements deal chiefly with details. In essentials our calendar promises to stand as a summary of earth history as well as a guide to times and formations investigated by men of rocks who pioneered in a new and growing West.

Geologic Time Chart

The following chart includes information about plants and animals of the past, as well as about rocks and major events of earth history. It should be read from the top (page 212) downward, in the order of geologic formations where they are not disturbed. Its lowermost division, the Cambrian Period, follows the Killarney Revolution, which appears at the top of the chart on page 201.

Although the Appalachian Revolution brought the Paleozoic Era to a close, so much of it took place during Permian times that it is placed in the Paleozoic. The Laramide Revolution, on the other hand, is so closely linked with Cenozoic conditions as to deserve a place in that era.

GEOLOGIC TIME CHART

CENOZOIC ERA

Periods, or Ages, Epochs and Their Length		Important Events	Rocks and Living Things	Dates
QUATERNARY PERIOD	Recent Epoch 15,000–25,000 yrs.	Ice Age glaciers melted, apparently for the last time. Extensive lakes formed. Lands remained high, with great differences in climate.	Many coarse-grained deposits settled in seas and on land. Man became the dominant animal while many large mammals died out. Insects increased abundantly.	15,000 25,000
	Pleistocene, or Ice Age 2,000,000 (?) yrs.	Four great glacial advances in Europe and in North America east of the Rockies; fewer in some other regions. Sierra Nevada and other mountains uplifted again.	Glacial drifts and outwash were spread widely; residual soils accumulated in non-glaciated regions. Many kinds of mammals spread widely and men came to Europe.	2,000,000
	Pliocene Epoch 11,000,000 yrs.	World-wide uplift and building of mountains continued, especially in western parts of North and South America. Great volcanic eruptions. Climates became cooler and more varied, resembling those of modern times.	Great thicknesses of volcanic rock formed in the West; loose, sandy material settled on coastal plains. Elephants, camels, horses and other mammals wandered from continent to continent.	13,000,000
	Miocene Epoch 18,000,000 yrs.	Early Sierra Nevada and Rocky Mountains were built. Eruptions began to form the Cascades and great lava plains of the West. Climates still mild and fairly uniform over great areas.	Volcanic rocks came to North America in the West. Elephants while apes appeared in the Old World and redwood forests grew near volcanoes of Colorado.	15,500,000 18,000,000

[212]

TERTIARY	Oligocene Epoch 10,000,000 yrs.	Erosion made most lands low, with mild and equable climates. Volcanoes erupted in the region of the Rockies and on the western Great Plains. Alps and Himalayas began to rise.	Volcanic formations in the Rockies; clay and ash deposits on the Great Plains. Hoofed mammals and carnivores became abundant. First members of the elephant family appeared in Egypt.	36,000,000
	Eocene Epoch 15,000,000 yrs.	Early Cenozoic mountains were eroded, piling clay and coarse sediment on lowlands. Climates apparently were mild; seas that overflowed continents were narrow.	Coarse sediments in narrow seas; volcanic rocks and oil shales in the West. Mammals became abundant, and early ancestors of the horse appeared in Wyoming.	57,000,000
	Paleocene Epoch 5,000,000 yrs.	Mountains still were high, probably growing higher during the early part of this epoch. Erosion was rapid; climates varied; seas upon continental areas were very narrow.	Conglomerates and sandstones in valleys; subbituminous coal in Montana and Wyoming. Primitive mammals became abundant; some were large and grotesque.	60,000,000

Laramide Revolution

Lands were raised throughout much of the world; the early Rockies appeared in a region of former swamps and seas. Climates became more rigorous and varied; the last dinosaurs died out, as had many other reptiles.

[213]

MESOZOIC ERA

Cretaceous, or Chalk Age 55,000,000 yrs.	Last great spread of shallow seas across continents, especially in the West. Land dominantly low, with great swamps; climates moderate and uniform.	Chalk formed in Europe and Kansas; coal in swamps of the West. Flowering plants appeared while horned dinosaurs evolved. Other reptiles flew or swam.	85,000,000
			110,000,000
Jurassic Period 40,000,000 yrs.	Lands eroded to moderate elevations. Shallow seas covered much of Europe and narrow basins in the West. Extensive swamps. Climates moderated greatly.	Sediments of many kinds settled on lands, in lakes and in seas. Plant-eating dinosaurs became very large. Birds appeared in Europe and mollusks were abundant.	123,000,000
			131,000,000
Triassic Period 35,000,000 yrs.	Continents still high; climates arid in many parts of the world. Late in the period there were volcanic eruptions and lava flows in the East.	Dinosaurs appeared and became common; mammals appeared. Many of the rocks are red or red-brown shales and sandstones, interbedded with gypsum and hardened lava.	165,000,000
			170,000,000
			180,000,000

PALEOZOIC ERA

Appalachian Revolution

Mountain building and glaciation of Permian times reached their greatest intensity.

Permian Period 30,000,000 yrs.	Continents became high; Appalachian, Ouachita and Ural mountains were built. Climates cooled, deserts developed and there was extensive glaciation, especially in South Africa and Australia.	Red beds, gypsum and hardened glacial deposits as well as lavas, limestones and other rocks. Reptiles became abundant; both they and amphibians grew large, with bony ornaments.	205,000,000
			225,000,000

Period	Description	Years
Carboniferous Period 60,000,000 yrs.	Much land was low and moist; seas spread and withdrew many times. Extensive coal swamps appeared in many parts of the world. Climates were equable and moist.	Much limestone formed early in Carboniferous times; then thick deposits of coal. Forests grew in many regions, reptiles appeared, and insects became common.	232,000,000
Devonian Period 40,000,000 yrs.	Shallow seas spread repeatedly; some uplift in the Appalachian region. Volcanic eruptions in New England, eastern Canada and the United States.	Limestones, shales and sandstones settled on sea bottoms; other sandstones piled up on deltas. Forests appeared; amphibians evolved from fish.	278,000,000 300,000,000 316,000,000
Silurian Period 30,000,000 yrs.	Seas became widespread; lands were low, with arid regions in eastern North America. Climates generally were warm and equable, except toward the end of the period.	Limestone, shale and other marine deposits became extensive; salt and gypsum formed thick beds. Land plants developed; sea scorpions and other marine creatures were abundant.	365,000,000 380,000,000
Ordovician Period 60,000,000 yrs.	Seas covered more than six tenths of North America; lands became low. Deltas formed in what now are New York and Pennsylvania.	Marine deposits of many kinds; some now are slate and marble. Marine animals of many kinds left vast numbers of fossils.	375,000,000 400,000,000
Cambrian Period 80,000,000 yrs. (or more)	Seas spread across North America three times. Lands were low and no mountains formed. Climates were mild and uniform.	Marine deposits in seas. Marine animals and seaweeds formed fossils in great abundance. Some became large, but others were tiny.	425,000,000 445,000,000 531,000,000

The only revolutions listed are those that closed eras.

[215]

CHAPTER XVIII

Railroads and New Surveys

On the eve of the 1860 election, crowds cheered as Colonel E. D. Baker appeared on a balcony in Sacramento. With Victorian oratory he urged his fellow Californians to support Abraham Lincoln, "the man the whole Northwest will cast its vote for." Why? Not merely, said the speaker, because Lincoln was a former rail splitter, a man of the people, a decent and honest citizen. No—and here Colonel Baker fairly thundered—the times demanded a man of vision who could plan a nation's destiny. Lincoln was such a man. Did he not believe in one United States? And did he not stand foursquare upon a platform which declared that "a railroad to the Pacific Ocean is imperatively demanded by the interests of the whole country," and that "the Federal Government ought to render immediate and effective aid to its construction"?

This idea of a railroad to the Pacific was by no means new. A Michigan editor had urged it back in 1832; by 1840 a wealthy New York merchant had dedicated his life to the project. Nine years later a Pacific Railroad Convention met in St. Louis, amid great enthusiasm but without visible results. In 1852 the California legislature invitingly granted the

Railroads and New Surveys

United States a right of way across the state "for railroad purposes." Congress replied with a grant of $150,000 for surveys of routes from the Mississippi Valley to the Pacific. The appropriation was renewed, providing funds for survey of several east-west routes and of others running inland but roughly parallel to the coast. Studies of the latter were completed in 1855–56.

Survey parties were under army command, for the government had no civilian service organized for exploration. Most of the parties, however, were accompanied by naturalists and geologists. Evans, who had collected fossils in the Badlands of South Dakota for Owen, crossed the Rockies and Cascades to Puget Sound; J. S. Newberry, who became professor at Columbia University, went from the Columbia River to San Francisco, passing ice-capped Mount Shasta without noticing its glaciers. A physician and chemist named Antisell recognized late upheavals in the Coast Ranges but tried to explain the Gila Valley as a crack or perhaps a narrow depression into which the river flowed. "Some such catastrophe must have occurred," he concluded, "for it is scarcely probable that the river unaided could have cut through such lofty hills as it appears to have done in its passage through these mountains."

This, of course, harked back to Von Buch; nor did other geologic results of the surveys advance earth science greatly. But the reports appeared in thirteen quarto volumes, some filled with costly colored plates which showed birds, beasts and Indians against backgrounds of more than realistic scenery. Printed in large editions, many copies provided country children with pictures to be cut out on rainy afternoons. Others went into schoolrooms, the few public libraries, and the bookcases of influential men. Lincoln and Herndon kept a full set, with the flyleaves inscribed "A. Lincoln from the Hon. F. E. Spinner."

One youngster who may have been thrilled by those reports was fourteen-year-old Clarence King. The son of a China

The Story of the Great Geologists

trader who had died at Amoy, young King had seen his family pass through bankruptcy when a steamer bearing large sums of money sank during a storm. In spite of this the boy was well educated at Newport and in Hartford's endowed high

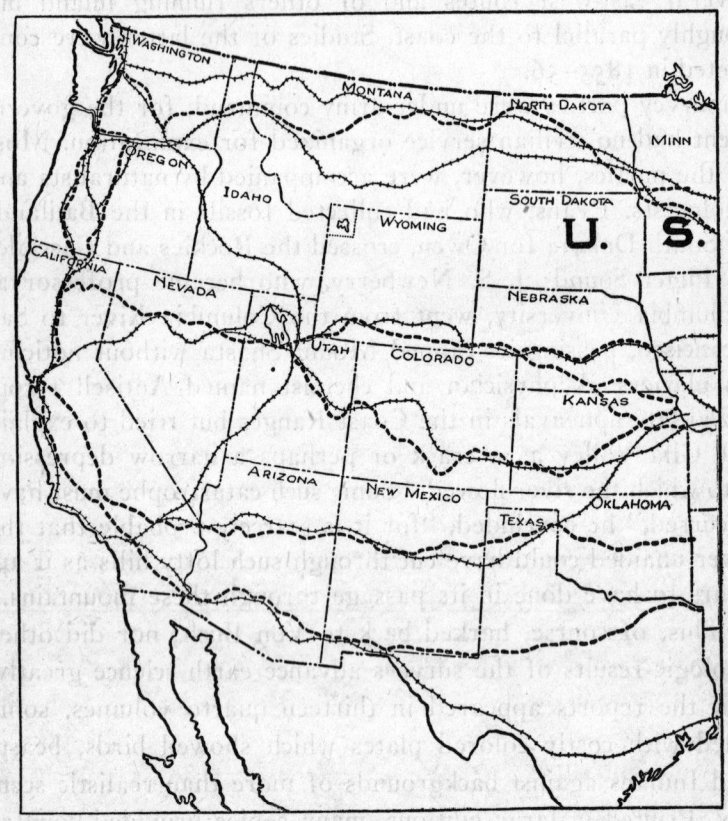

Principal routes of the Pacific Railroad surveys of 1853–56 in heavy broken lines. Note how closely they are paralleled by modern railroads.

school, with emphasis on classic and modern languages. Summers were spent fishing, hunting and collecting plants in the Green Mountains—such a thorough balance for languages that King entered Yale in 1859 as a scientific student. He specialized in geology and mineralogy under James D. Dana

Railroads and New Surveys

and George J. Brush, then the foremost teachers of those subjects in North America.

King graduated in 1862, studied glaciation under Agassiz, and became devoted to Ruskin's ideas of art and artistic literature. But he had also developed a desire to know the West, and when a friend's health failed the two young men planned a trip to the Pacific coast. They set out in May 1863, taking trains as far as St. Joseph, Missouri, where rails came to an end. There they joined a well-to-do emigrant family whose children King had amused during the tedious, dirty train ride. They followed Frémont's route up the North Platte and down the Humboldt in Nevada, with the emigrant wagons moving so slowly that King and his friend had plenty of time for saddle trips into near-by country. On these they examined rocks and plants, with mild adventures in the form of escapes from unfriendly but not too warlike Indians.

King and his friend left the party in western Nevada to visit mines of the famous Comstock Lode. In Virginia City they rented lodgings; that night the rooming house burned, and with it went the young men's money as well as most of their clothes. Since the friend still was unwell, King worked in a quartz mill to pay board and earn a few extra dollars. With just enough for necessities, the pair walked across the Sierra Nevada and to Sacramento, where they took a river boat for San Francisco. There King got money from home and joined the Geological Survey of California as an unpaid assistant. For three years he did exploratory work under J. D. Whitney, one of the men who had surveyed northern Michigan while Owen was in the Chippewa country. King discovered fossils in rocks of the rich Mother Lode, studied volcanoes of the southern Cascades, and explored the country about Mount Whitney, which he named for his superior. The winter of 1865–66 was spent in the deserts of southern California, amid hardship and considerable danger.

California had been pleased when the railroad surveys were made, but bitter when they led only to those thirteen quarto

The Story of the Great Geologists

reports. During the War of Secession there was fear that this state, too, would leave the Union to become an independent nation. To prevent this Congress made good the Republican campaign promise; a railroad was subsidized and pushed forward even though both men and money were needed to fight the South. As peace came King perceived another need; Western mineral resources should be developed as a means of building local industries which the railroad might serve. Those industries would be scattered along the 1260 miles of line west of Cheyenne, attracting workmen, establishing towns, and bringing in capital. They would make the West an integral part of the country, rather than a wilderness across which steel rails should lead to California.

Those were days of the pioneer prospector, that bearded man with rifle, pack horse and pick who sought ores (chiefly of gold and silver) by perseverance and rule of thumb. This method might succeed, as California showed, but King knew that sound development called for information which prospectors could not secure. It must come through a geologic survey, not in one state or another, but through a belt along the whole Overland Route from Colorado to San Francisco.

His plan made, King left California to spend the winter of 1866–67 as a lobbyist in Washington. He received no salary and paid his own way, but was rewarded when Congress passed the bill which he probably wrote. For King himself—then barely twenty-five—became director of the new United States Geological Exploration of the Fortieth Parallel.

The survey nominally was under the Army's chief of engineers; actually it was independent, with King in almost autocratic charge. He left Washington as soon as he could and, since the railroad still was far from complete, took ship to Panama, rode across the isthmus, and sailed to San Francisco. There three months were spent organizing and outfitting "the best-equipped party ever to enter American field geology." August came before work could begin at the eastern foot of the Sierras, along the Comstock Lode.

Railroads and New Surveys

Here was no new deposit to be developed: the Lode, with its mines a thousand feet deep, had produced a hundred million dollars' worth of silver and was threatening the nation's monetary system. Yet it made an excellent place to start, and by 1870 enabled King's survey to complete a report on mining industry which became the one authoritative manual of precious-metal mining and metallurgy in North America. Meanwhile the field party crossed Nevada and western Utah to Great Salt Lake, worked eastward to the Uinta Mountains, and built up large collections for study. Thus it explored country now famous for deposits of silver, copper, manganese and other metals whose importance increases each year.

Field work, finished in 1869, used all the time King had planned for this part of his survey. He therefore took his staff to New Haven, where Yale University offered its facilities for study of specimens and preparation of reports. Next summer, however, Congress suddenly extended the survey four more years, and a telegram from Washington sent the party into the field. Since it was too late for work in the Wasatch Range, King spent the season on Mount Shasta and other old volcanoes of the Cascades. There he discovered few ores but found the first glaciers known within the United States, at the same time explaining gravel mounds and kettle holes of faraway New England.

Two years later King achieved public fame by exposing a mining fraud. It began with seemingly honest announcements of a diamond deposit in southern Arizona, so great as to affect the world's market for those precious gems. Though the exact locality was secret, King and his aides recognized it as part of a region they had explored. Off they went to get more information, only to find that the mine had been very crudely "salted." King rode day and night to San Francisco, burst into the diamond company's office, and demanded that sales of stock cease at once. When officers suggested that a delay would be suitably rewarded, he replied, "There is not enough money in the Bank of California to induce me to delay this announcement a single hour!" Given banner headlines in the newspapers, King's report saved would-be investors from grave financial loss.

By 1873 the Survey reached the Great Plains, crossing the bed of large lakes that once lay in Wyoming, Utah and Colorado. But 1877 came before the staff could study collections stored at Yale and write a series of massive reports. During this time an assistant was sent to Europe for the purpose of examining other national surveys, as well as to buy books and magazines not available in the United States. He also induced Professor Zirkel, world-famous petrographer of Leipzig, to cross the Atlantic and spend some months in a study of the Western collections. One result of this trip was a book on the microscopic make-up of rocks—a book which, though written by a German, revolutionized this phase of American geology.

King himself closed the Survey's reports with a volume of eight hundred large pages which summarized discoveries of the entire staff. In it he presented the first systematic account of Western formations, which begin with schists and granites of Archeozoic age and range through succeeding systems to deposits of modern times. More surprisingly, he found that those formations recorded features of the Pre-Cambrian continent resembling those of North America today. Thus mountain ranges of Laurentian and Huronian times occupied a north-south belt near the Pacific, while east of them lay a level region then covered by a shallow sea. In time these relationships were reversed, with a series of depressions and uplifts that alternated through age after age.

During one of those uplifts—a relatively late one—great volumes of hot granitic magma were forced into folded strata of the Sierra Nevada. At other times the Wasatches were tipped upward, the Rocky Mountain front appeared, and the Uintas arose as an oval fold which carried ancient marine strata to amazing altitudes. King was enough of a catastrophist to believe that folding proceeded with such speed as to far outstrip erosion. As a result the top of the dome formed blunt mountains five to eight miles in height.

Success of work along the fortieth parallel inspired other federal surveys. One of these was commanded by an army engineer named Wheeler, who knew next to nothing of earth science, and whose chief geologist was discouraged by military methods and delays. Another was headed by a brusque but able ex-major, while a third was directed by a civilian whose devotion to both geology and the West was even greater than King's. He built up a large and useful organization which failed only when all three surveys descended to harmful rivalry.

Ferdinand Hayden was thirteen years older than King, having been born at Westfield, Massachusetts, in 1829. His was no background of culture and former wealth, for the elder Hayden was a poor man who died when Ferdinand was ten. Unhappy under a stepfather's rule, the boy soon went to live with an uncle on an Ohio farm. At sixteen young Hayden taught country school for a term, refused his uncle's offer of adoption, and decided to give up life as a farmer. Walking to Oberlin College, he enrolled in the preparatory department and began to earn his way through school. Without funds from either mother or uncle, he managed to meet all expenses and graduated in 1850 with a master's degree. He then tramped and caught rides to Albany, New York, and in three years earned an M.D. from the Albany Medical College. He also became a friend of James Hall, who insisted that no man so plainly cut out for science should become a damned pill-peddling sawbones.

Though Owen's great report was unpublished, geologists already were aroused by stories of fossils which Evans had found among the barren, pinnacled Badlands and in shale banks a few miles away. Hall cared nothing about the bones but was eager to study the petrified shells which showed that those shales had settled during Cretaceous times. Would Hayden go to get them if his expenses were paid? No wages, for

The Story of the Great Geologists

Hall couldn't afford them—and besides, the trip itself should be good pay.

Hayden accepted the offer and started out with one of Hall's regular assistants. In St. Louis they encountered Evans and with him organized an expedition that took boats up the Missouri and wagons across rolling plains to the Badlands. Collecting was good, and by autumn Hayden was back in Albany with fossil shells for Hall to examine and bones for Dr. Joseph Leidy, who had described Evans' original fossils. When these specimens were delivered Hayden turned westward again, determined to carry on geologic exploration of what then was Nebraska Territory.

Such work now is done by members of official surveys, or by parties sent out by museums with money for exploration and research. Hayden, however, had no such backing, nor did he have money of his own with which to pay expenses. Unwilling to give up, he appealed to officials of John Jacob Astor's American Fur Company, who allowed him to travel free with its parties. For almost two years Hayden journeyed up and down the Missouri between Council Bluffs and Fort Benton, and from Fort Union up the Yellowstone as far as the mouth of the Big Horn. He rode on both paddle steamers and keelboats, going ashore while flotillas tied up at trading posts, or examining rocks and collecting fossils while boatmen struggled with rapids or bars. To earn money he did odd jobs for traders, who often were glad for a "hand" to arrange their stocks, cast up accounts, or pack bales of pelts for shipment.

By 1856 Hayden was back in St. Louis, where Lieutenant G. K. Warren asked him to serve as guide and naturalist on an expedition to the Sioux country. Hayden first prepared a preliminary report from his own observations and then set out in May with Warren's party, which traversed and roughly surveyed the Missouri from "Fort Pierre to a point 60 miles north of the mouth of the Yellowstone." Then came a season of official work in the Black Hills and the Territory of Kansas, followed by an expedition to the "country through which flow

Railroads and New Surveys

the principal tributaries of the Yellowstone River, and the mountains in which they and the Gallatin and Madison forks of the Missouri have their source." The party, led by Captain W. F. Raynolds, left St. Louis in May of 1859, took steamboat to Fort Pierre, and then traveled overland. After reaching the Yellowstone River in north-central Montana it detoured far southward, wintering near the North Platte. Exploration was resumed in May 1860, along routes now popular with western tourists. The Missouri was reached at Fort Benton in July, and in early October the party disbanded at Omaha.

On these trips Hayden discovered "Primordial," or Cambrian, rocks near the source of the Yellowstone and described them as having been changed by heat that apparently came from beneath. He identified formations of Carboniferous and later ages, divided Tertiary deposits, and concluded that the region of the Badlands had once enjoyed a climate as mild as that of the modern Gulf States. He also determined that the Rockies began with gentle arching near the end of Cretaceous times, now known to have been some 60,000,000 years ago. Great disturbances, he said, came much later, at a time when faults divided the uplands into precipitous blocks.

The War of Secession halted Hayden's explorations, for he promptly joined the Army as surgeon with a division of cavalry. Mustered out in 1865, he was appointed professor of geology and mineralogy at the University of Pennsylvania and returned to the Badlands in 1866 with an expedition sponsored by the Academy of Natural Sciences in Philadelphia. Leaving Fort Randall, South Dakota, on August 3, Hayden traveled six hundred and fifty miles with a six-mule team and brought back a fine collection of fossils. The paleontologist, Leidy, again described them in a work of four hundred and fifty pages and thirty lithographed plates.

Nebraska became a state in 1867, and Congress directed that five thousand dollars of unspent territorial funds be used for a geological survey. Next year the grant was repeated, and

The Story of the Great Geologists

Hayden was ordered to work westward into the Territory of Wyoming. In 1869 he took charge of a newly established Geological and Geographical Survey of the Territories, under the Department of the Interior. Investigation began with a reconnaissance along the Rocky Mountain front from Cheyenne to Santa Fe, but between 1870 and 1872 work was extended to the Laramie Range, Yellowstone Park and the Snake River Basin. Colorado was added in 1873; by 1877 Hayden had three parties widely scattered in that state and two others that worked in Wyoming, Utah and Idaho. In 1878, with appropriations reduced by rivalry, he still was able to send expeditions along the Yellowstone, Snake and Wind rivers of Wyoming and Idaho.

The Survey was geographic as well as geologic, and geography had not come of age in the 1870s. To some people it meant little except boundaries and place names; to others it signified maps with key points accurately located and shading to indicate slopes. For Hayden and most other officials, it implied a wide range of historical, descriptive and scientific data, intended to give an all-round picture of regions selected for exploration. Hayden's staff, therefore, included botanists, zoologists, archeologists and army surgeons who studied rodents or birds. Extending geology to include fossils, he engaged a Boston scientist to describe ancient insects from Colorado, while Cope, a brilliant Philadelphian, wrote a thousand-page treatise on mammals. Leidy prepared one volume on this subject, too, though his major work dealt with one-celled living creatures found in ponds and streams.

Scientific exploration, however, was not concentrated under Hayden and King. The War Department had directed railroad surveys of the 1850s, had sent Captain Stansbury to Great Salt Lake, and had dispatched Lieutenant Ives to explore the Colorado River in an iron stern-wheel boat. In 1869 it authorized Lieutenant G. M. Wheeler to begin topographic work of a military nature in western Utah and Nevada. By 1871, however, his organization was established as a survey

Railroads and New Surveys

of regions west of the one hundredth meridian, at first with one geologist. This number in time was increased to three, with a mineralogist besides. Publications of the Wheeler Survey dealt with geology and mineralogy of six Western states, as well as with plants, insects, fish, birds, mammals and

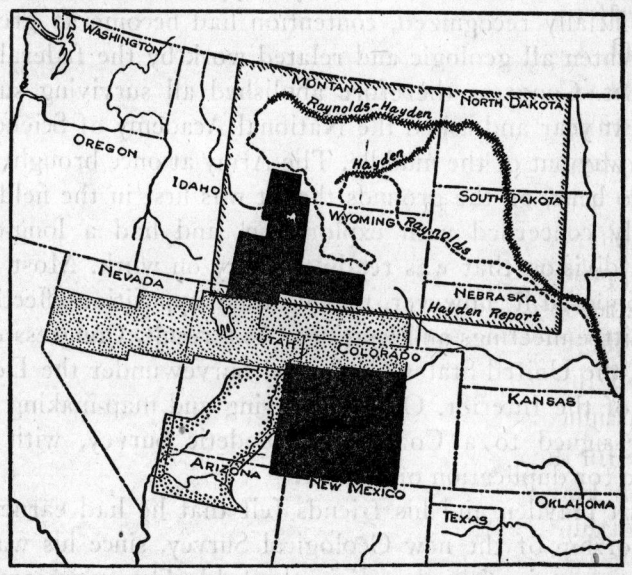

EXPLORATIONS OF EARLY SURVEYS

Areas explored by the King survey (fine dots), Powell survey (coarse dots) and Hayden survey (black). The large rectangle shows the region included in Hayden's report on the Raynolds expedition; the broken and crosshatched lines show Raynolds' route and Hayden's travels with the American Fur Company.

archeology. A Bowdoin College professor wrote on invertebrate fossils, while Cope described ancient vertebrates.

In 1879 still another survey was added, to deal with the Rocky Mountain region. It, too, was geographic as well as geologic, publishing eight important volumes on Indians of areas as far away as Washington and Alaska. In the earth sciences it dealt with the Uinta Mountains, the Black Hills, plateaus and mountain ranges in Utah, and arid lands in many parts of the Southwest.

These surveys were bound to overlap, and from their duplication came conflict. Each sought priority in publication; each tried to get its own funds increased, if necessary by reducing those available for its rivals. By 1877, when the last survey was officially recognized, contention had become so great as to threaten all geologic and related work by the federal government. Congress therefore abolished all surviving surveys the next year and asked the National Academy of Sciences to find a way out of the muddle. The Army at once brought pressure to bear, on the grounds that it was first in the field, was directly concerned with exploration, and had a long-established division that was ready to carry on work. Most scientists opposed it, however, making their opposition effective in committee meetings and hearings. As a result, Congress established the United States Geological Survey under the Department of the Interior. Other surveying and map-making work was assigned to a Coast and Geodetic Survey, with little chance for duplication or conflict.

Both Hayden and his friends felt that he had earned the directorship of the new Geological Survey, since his was the largest organization already in the field. They were amazed, therefore, and disappointed when the position was offered to King, who frankly did not want it and was unwilling to hold it long. Yet Hayden refused to raise new quarrels by asserting his claims, and in 1880 was appointed geologist under his former rival. For six years he continued work from his home in Philadelphia, where he had the advantage of libraries better than any then in Washington. His service was ended by alarming symptoms of locomotor ataxia, which made such progress that he had to resign in 1886. Death came three days before Christmas in 1887.

King left the new Geological Survey five years before Hayden resigned. He had become director under protest, agreeing to stay just long enough to organize work, select a staff, and

Railroads and New Surveys

guide it into operation. All this he accomplished within a year, giving emphasis to practical studies yet establishing innovations of sound scientific value. Chief of these was a laboratory in which the exact methods of physics, mathematics and chemistry were applied to the study of rocks. Though too technical to interest most miners, such studies would add enormously to knowledge of the earth.

Resigning in 1881, King spent the rest of his life as a mining engineer. He located deposits, planned workings, supervised them; when his clients encountered legal problems he went with them into court. In most suits he acted both as scientist and interpreter of mining law, assuming full control of the case. Each fact of importance he verified, building up such masses of evidence that few really able lawyers attempted to shake his testimony. With a reputation for both knowledge and integrity, his informal reports often settled conflicts which in other hands might have dragged through years of legal altercation.

No booster or politician of science, King seldom attended meetings or contributed to proceedings of learned societies. He preferred to work out of doors; when tired he sought recreation with literary and artistic friends. They shared his joy in good pictures, fine sculpture, great poems or prose; they understood, as some scientists did not, the love for good writing which made him keep imperfect manuscripts from publication. Those that did appear were planned in advance to the final detail and then were written rapidly with a minimum of correction. None rivaled the *Systematic Geology* of the Fortieth Parallel in bulk, and none quite met King's own standards. Yet a volume on mountaineering in the Sierra Nevada became popular in England and still stands as one of the best-written books of scientific travel in the West.

As a mining man, King became acquainted with deep-seated phenomena of rocks, which led him to consider problems of uplift and sinking, as well as the age of the earth. In the latter he relied partly on physical studies by men whom he had appointed to the United States Geological Survey. Using ideas of cooling developed by Kelvin, of England, King made elaborate mathematical computations which seemed to show that there was "no warrant for extending the earth's age beyond 24 millions of years." Modern data, also physical, place it nearer to a hundred and fifty times that sum.

King also protested against the extreme uniformitarianism of some who followed Lyell. Earth's present, of course, was the key to its past; depression, uplift and erosion have gone on in every geologic age. But their rate of operation has varied; at some times it was moderate but at others it was rapid. Tracing the roots of mountains long since worn away, King concluded that "the harmless indestructive rate of geologic change of today" could not be "prolonged backward into the deep past." In this he was both right and wrong, for although some geologic agents undoubtedly have worked more rapidly than they do today, others have operated more slowly. And when a single river, the Mississippi, can carry 5,777,000,000 tons of rock from its basin in a year, it is neither harmless nor "indestructive."

In dealing with hot magmas that welled up from great depths, King pictured a gradual change as the doughy stuff moved and solidified. At the outset, said he, a large mass might form granite containing an abundance of light-colored quartz and feldspar, but few dark minerals. The latter somehow remained molten, working out to edges of the intrusion, where they entered weak zones or cracks. There they might harden in sheetlike sills and dikes, or they might go all the way to the surface and appear as blackish lava flows. Such leftovers seemed to account for vast basaltic plains of the Snake and Columbia basins, as well as volcanoes that ranged from Alaska southward through Mexico. The process of differentiation itself helped to explain mineral deposits at the edge of great granitic masses, as well as in dikes and veins.

King also placed great reliance on pressure as a means of changing rocks. Shale, squeezed and crumpled, turned into

Railroads and New Surveys

slate as belts of land were folded into mountain ranges. Still more intense squeezing produced schist, the platy, intensely crumpled rock whose layers often are covered with gleaming mica. Further compression turned schist into gneiss, and finally into crystalline granite. Though this series itself probably was wrong, its author made a great advance when he announced that granite could be a metamorphic rock as well as a hardened magma.

A stocky, bearded man of great endurance, King also was subject to sudden breakdowns in which he became intensely nervous and suffered amazingly. In 1900 he went to the Klondike, where he apparently contracted tuberculosis during another collapse. Next year he fell ill of pneumonia; though he recovered, the other disease was revived and progressed at a rapid rate. Several changes of climate brought no relief, and in December of 1901 he died at Phoenix, Arizona. Young men to whom his work along the Fortieth Parallel had been legend were amazed to learn that he failed by two weeks of reaching his sixtieth birthday.

CHAPTER XIX

Canyon's Conqueror

IT HAS BEEN SAID that geology discovered the West through Hayden and King, but that is an exaggeration. Other men reached the Rockies and the Pacific before them; other men shared the task of exploring plains, deserts, canyons and plateaus. Among these scientific pioneers was a one-armed former soldier and professor who rivaled Hayden in his devotion to geology and outranked King as a scholar. An able leader and executive as well, he became director of the federal Geological Survey when it settled down as a going concern in 1881.

John Wesley Powell was born in 1834 at Mount Morris, in western New York. His father was a Methodist minister who moved to southern Ohio when John was only four years old. There the elder Powell's fiery abolitionism caused such resentment that his son was abused by other boys and had to leave the public school for one taught gratuitously by a prosperous elderly neighbor. Much instruction was carried on out of doors, to John's delight and the disapproval of townsfolk who thought Old Man Crookham a fool.

In 1846 the Powells' possessions were loaded into a wagon

Canyon's Conqueror

and two four-wheeled carriages, one of which twelve-year-old John drove over dusty roads to Wisconsin. There his father combined farming with circuit riding, which meant preaching at one church after another in a regular series, or circuit. While he was away John took care of the farm and, a dozen times a year, hauled loads of produce to the nearest market. Each round trip lasted five or six days—five if all went well, but six if muddy roads delayed the loaded wagon.

In 1852 young Powell began to teach in a one-room school where half the students were older than their instructor. For nine years he alternated terms of teaching with trips afield and study at varied colleges in Ohio and Illinois. His scientific work began in 1854 with systematic collection of mollusks found in prairie lakes and streams. Succeeding years saw him drifting down the Mississippi or gathering shells from the Ohio, the Illinois, the Des Moines and a series of lesser streams. He became secretary of the Illinois Society of Natural History, and in 1860 visited the South on a lecture tour. Sampling pro-slavery sentiment in the light of his own abolitionism, he concluded that only war could solve the conflict. When war came, he enlisted in May of 1861, went to the front as sergeant major, and soon received a second lieutenant's commission. A smattering of engineering aided his rise, for he was able to lay out roads, design bridges, and plan entrenchments or camps.

During the winter of 1861–62 Powell recruited a company of artillery and therefore was made its captain. In March a short leave allowed him to visit Detroit, where he hastily married a cousin, Emma Dean, to whom he had been engaged. Instead of parting, the young couple returned to the battle zone together. There Mrs. Powell served as a voluntary nurse during the Battle of Shiloh, when a rifle ball struck her husband's right wrist, glancing toward the elbow. Hasty surgery added to the injury, demanding a second operation which reduced the forearm to a stump. This, too, was done so poorly that for thirty-two years Powell suffered repeated and increas-

The Story of the Great Geologists

ing pain. Relief did not come from a third operation until 1894.

As soon as Powell's first wound was healed he returned to service and as major assumed command of sixteen field batteries. "He loved the scarlet facings of the artillery," wrote an associate, "and there was something in the ranking of batteries and the power of cannon that was akin to the workings of his own mind." He earned Grant's admiration, though lesser officers sometimes smiled at a commander who gathered mosses while deploying in woods and collected fossils when his men dug them from trenches before Vicksburg.

Mustered out in the summer of 1865, Major Powell refused a political job for the professorship of geology at Illinois Wesleyan College. Two years later he went to the State Normal University with a salary of fifteen hundred dollars and a thousand-dollar appropriation yearly to build up that institution's museum. He took classes out of doors for instruction, led public discussions of science, and urged legislators to improve scientific instruction in the state. During the summer of 1867 he made educational history by taking sixteen "naturalists, students and amateurs" to the Rocky Mountains of Colorado for field work and exploration. They worked along the Front Ranges, climbed Pikes Peak, went westward to Middle and South parks, and collected thousands of specimens for shipment back to Illinois.

In Middle Park Powell camped near a trading post run by Jack Sumner, another Union veteran. Together they studied color phases of bears, visited hot springs, and discussed fields for scientific exploration. Powell told of his plan to visit the Badlands; a plan frustrated by hostility of the Sioux, but one which might be followed in 1868 if Sumner would act as guide. The trapper-trader shrugged and spat his refusal. The Badlands, he opined, were stale—a region known since the early days of the fur trade and familiar to scientists since that fellow Owen had written his report. Why not do some honest-to-God

Canyon's Conqueror

exploring? Why not go down the Colorado River, where even Indians had not been?

That scheme was foolish, suicidal, foredoomed to failure; all this and more Powell asserted while making up his mind to go. Sumner supposed they would start in 1868, and was disgusted when Powell came back from Illinois with his wife, brother and two dozen students "about as fit for the Colorado as hell for a powder house." They collected, climbed Longs Peak, and then went home while the Powells, Sumner and several trappers made a winter camp on the White River of northwestern Colorado. Powell journeyed down the Green, the Grand and the Yampa, and for the first time studied ways and languages of Indians. Since he had no formal training in ethnology, he apparently began these studies from a natural interest in red men.

Winter evenings were spent making plans, and when spring came Powell went to Washington in search of government support. Urged by Grant, Congress decided that rations might be drawn from military posts, while the Army provided barometers, sextants and other instruments. The Illinois State Normal University agreed that Powell's salary and museum fund should be used, another state school gave five hundred dollars, the Chicago Academy of Sciences contributed something, and friends gave as much as they could. Four sturdy boats were built in Chicago and taken to Green River, Wyoming, then a straggling town of adobe and canvas beside the Union Pacific Railway.

There, about the twelfth of May, Powell greeted the seven Westerners who formed the backbone of his staff. Accustomed to light travel and hardships, they noticed that the major had brought box on box of "necessary trinkets," as well as a "young scientific duck who was not at all necessary." He "vamoosed camp that night," Sumner later recalled, urged eastward by condescending smiles, coarse sand in his food, and a filthy sock that seemed to come, dripping, from the coffeepot.

While waiting, Sumner and his cronies had painted Green

The Story of the Great Geologists

River red, gambling, fighting and drinking whisky as fast as barkeepers could dilute it with water and alcohol. Now they settled down to work, packed the boats for an early start, and waited impatiently while Powell shifted his varied cargo of guns, ammunition, instruments, and food enough to last ten months in case accident or ice should delay the journey. "Ice!" snapped a trapper as he repacked. "Ice when Injuns say that there Colorado runs a mile down under hot ground! But then, all hell may freeze over if we stay here long enough!"

On May 24 the party pushed off while gamblers offered odds that not one of its men would return. Their course led down the Green River, past buttes and slopes of laminated shale laid down in a Tertiary lake. The major's brother sang "Flow Gently, Sweet Afton"; mountain men cursed as they broke new oars and roared out laughter when an ex-bull-whacker complained that his boat wouldn't gee, haw nor whoa worth a damn. Red Canyon provided rapids and thrills; Lodore, where the Green River cuts the Uintas, had a falls that smashed one boat to bits and stranded its occupants on an island. They were rescued—and then denounced by Powell, who charged them with carelessness. Why hadn't they followed his signal to land? And how dare they try to excuse themselves by saying that no signal was given?

Currents slackened in northeastern Utah, where the river now flows between irrigated fields near the settlement of Jensen. Then Desolation Canyon closed in, with ninety-seven miles of swift water where Powell's own boat overturned, with loss of guns, instruments and food, as well as bedding rolls. A wide valley lay before the Book Cliffs; then came canyons that grew deeper and deeper till the Green River joined the Colorado (there and then called the Grand) between cliffs thirteen hundred feet high. The party climbed out to view the country above, threw away two hundred pounds of spoiled food, and sieved their oft-moistened flour to get rid of solid lumps and mold. Barely two months' provisions remained when the boats were pushed off from a crumbling sandbank into the Colorado.

Canyon's Conqueror

Here canyons deepened and rapids grew still more swift, forcing the men to portage or let boats downstream with ropes while Powell examined rocks or checked locations and elevations with battered instruments. He drove both himself and his staff, worried when barometers broke, and flew into purple-veined rage when Bill Dunn soaked the last usable watch. The result was an ugly quarrel in which the younger Powell was challenged to a duel, Sumner gave the major a tongue-lashing, and Dunn was ordered to leave the party as soon as the canyon walls would permit.

This outburst can be blamed on frayed nerves; on worry, exhaustion, lack of food, and Powell's agonizing arm. Yet it cast gloom over boats and camps along almost two hundred miles of the Grand Canyon. Bright Angel Creek gave a little relief, though the men used their time to rip out oars while Powell examined schists, granites, and ruins of Indian houses. There the last dampened baking soda was lost, so that men already half starved had no food except coffee, dried apples and doughgods made from water and musty flour. On August 27 they reached a roaring fall, made by boulders washed into the Colorado from canyons of tributary streams. There Bill Dunn and two other men left the party—Dunn because he was ordered to go, the others because they were Bill's friends and because they felt that the major had nagged them ever since the wreck of their boat. Powell himself once decided to leave the river, but changed his mind when the party's youngest members vowed they would go on alone. At last six men got into two of the boats, swung out into the current, and went through the falls without harm. Next day they passed the mouth of the canyon and on August 30 were greeted by Mormons at the mouth of the Virgin River.

There camp was made while a bishop brought food: melons and other luxuries which the party had not tasted for months. Major Powell and his brother set out for Salt Lake City, but the others went on down the river until two—Jack Sumner and the cook—finally reached tidewater. Meanwhile the men

The Story of the Great Geologists

who left at the falls climbed to the plateau and there were killed by Shevwits tribesmen who mistook them for miners who had raped and then shot three squaws. The story came

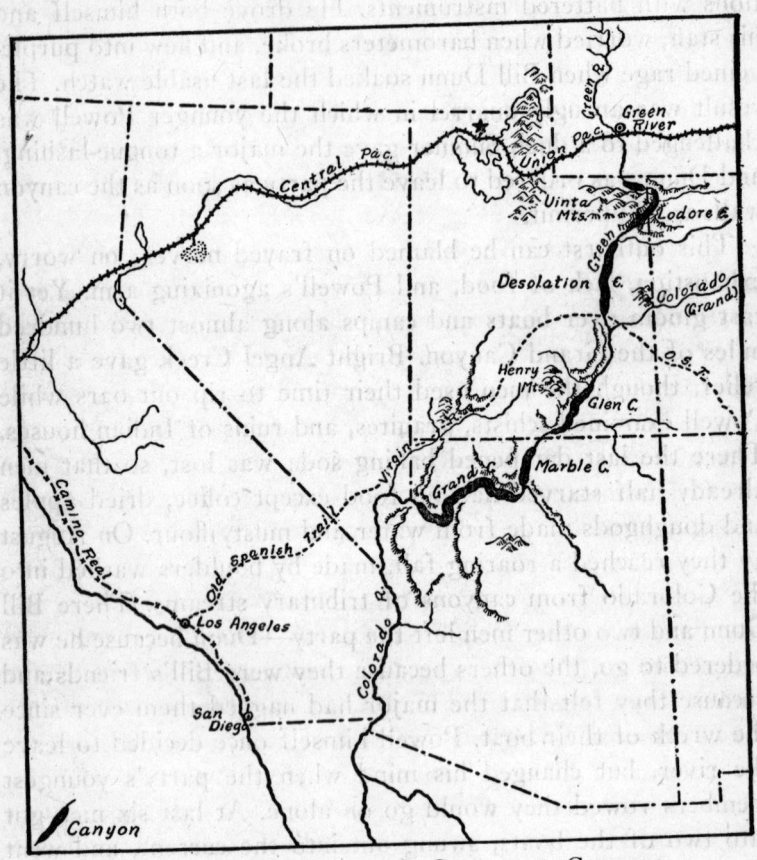

POWELL'S ROUTE DOWN THE GREEN AND COLORADO

Showing the major canyons, some early trails and the first transcontinental railway. Its two parts joined near the star north of Great Salt Lake.

out a year later when the major, then known as *Ka pu-rats,* or "Arm-off," explored Arizona's plateaus.

Five years passed before Powell told of his work in a report on *Exploration of the Colorado River of the West,* published

Canyon's Conqueror

by the Smithsonian Institution. During this time he revisited the river thrice: in 1870 to explore uplands and arrange for supplies, and in 1871-72 with three sturdy boats that again set out from Green River and drifted slowly downstream. This trip, with scientific and photographic aides, provided facts to replace the sketchy data secured in 1869.

In his report, a lively non-technical volume, Powell combined diaries of the two boat trips and made no mention of disagreements during 1869. Unintentionally, perhaps, this condensed version implied that Bill Dunn and his friends left the canyon solely through fear of the rapids. Unfortunately, too, imaginative woodcuts greatly exaggerated the fury and narrowness of the stream, while others showed events of 1871 as if they had happened two years before. Decades passed before some of these lapses were corrected by an author who had surveyed a possible railway route through the canyon and had consulted men who were with Powell in 1869. One of these was Jack Sumner and another was Hawkins, the youthful cook whose determination had encouraged the major to complete his original trip.

For a while Powell served under Hayden, having charge of one division of the Territorial Survey with almost complete independence. It was this division which in 1877 became the United States Geological and Geographical Survey of the Rocky Mountain Region. Powell had general charge, but devoted most of his own attention to arid lands and Indians. While the King Survey settled down to write reports, Powell's was one of the three organizations which in 1878 fought to a finish for federal approval and support. The fiery major tried hard to reach some form of agreement; when that effort failed he frankly advised the suppression of all existing surveys and establishment of a new one. He apparently pulled the wires which sent Congress to the National Academy of Sciences for advice, and the plan accepted was essentially his even though King was consulted and gave his full approval.

In spite of this, Powell declined directorship of the new

The Story of the Great Geologists

United States Geological Survey, preferring to head a bureau for the study of American Indians. But when King resigned in 1881, legislators and scientists alike insisted that the major head both organizations.

The new Survey had hired several men from its predecessors; Powell re-engaged others and also gave part-time appointments to professors in colleges. Standards sometimes were low, for trained men were scarce, yet politicians found the Survey no ready refuge for deserving relatives and friends. "Senators' nephews" who did get jobs found themselves doing camp work or acting as rodmen for surveyors. Powell insisted that the duty of a scientific administrator was to plan vital investigations and put them in the hands of the best men to be found. "If, then, improper persons are employed, it is wholly the Director's fault."

Such frankness impressed legislators, as did the scope and practical value of studies undertaken by the Survey. It had begun with an annual appropriation of $106,000 and thirty-nine employees; within eleven years it was receiving $801,000, while its staff formed the largest scientific body anywhere in the world. By congressional action its work had extended beyond the public domain to the entire country, and included the making of topographic maps as well as the study of surface deposits and deeply buried formations. In designing the maps, Powell adopted a system of brown lines connecting points of equal elevation, with a given height, or contour interval, between successive lines. Other symbols represented streams, lakes and swamps, as well as human structures that ranged from houses to breakwaters and roads. With simplicity yet precision and detail they showed what each region was like, how it was being used, and what problems it offered to builders of cities, railways or roads.

The need for such maps was plain, but detailed technical reports were less clearly useful yet more costly to prepare and publish. To insure support for them, Powell kept friendly legislators informed and argued long though tactfully with

Canyon's Conqueror

those who were skeptical. He urged that thorough reports were essential to the Survey and that they demanded continuous support. Only with it could men devote themselves to fundamental problems, without fear of reduced wages, dismissal or transfer to less significant work.

Essential, too, was an esprit de corps; that complex of loyalty, satisfaction and pride which we now group under the one word "morale." No longer the worried, disappointed leader, Powell encouraged his men by showing that he believed in them and their work, and by doing all he could to assist them. He gladly hired better scientists than himself and took care never to plague them with unnecessary rules. As an administrator he was genial, direct and informal, humming as he worked at his desk and going to chat with staff members instead of sending memoranda. When they faced problems he was free with his own ideas. "The major's generous," one assistant remarked. "He'll give any man the best thought that ever enters his head!"

Like King, Powell was eager for the West to develop on a sound and realistic basis. He had argued against the "Great American Desert" when that fallacious term spread across maps from central Prairies to the Sierras and Cascades. During the 1870s and 1880s came a reversal, and Powell found himself forced to combat an even more fallacious assertion that no true deserts were to be found in the West. Moist seasons gave support to this new error, which coincided with a wave of emigration and was spread by clever advertising. Even experienced army men, who should have known better, declared that railroads, cultivation and telegraph lines had permanently changed the climate. Credulous settlers believed them and bought land in regions where normal rainfall meant utter failure of crops.

For this folly Powell proposed an antidote in a commission to appraise physical, climatic and economic resources of dry lands in the West. He gave it two years of hard work, with additional field studies by his ablest geologist. Results com-

The Story of the Great Geologists

prised a *Report on the Lands of the Arid Regions of the United States*, for which three of Powell's associates also provided chapters. It showed that man could not readily change climates, that much of the West could never be farmed, and that irrigation was the best guarantee of protection in regions where precipitation was less than twenty-eight inches per year. But irrigation demanded careful study of streams and the building of dams for reservoirs to save water from winter rains or snows. Nothing, said he, should be done in a hurry, nor should water be divorced from land. "If, in the eagerness for present development, a land-and-water system shall grow up in which the practical control of agriculture shall fall into the hands of water companies, evils will result therefrom that generations may not be able to correct, and the very men who are now lauded as benefactors to the country will, in the ungovernable reaction which is sure to come, be denounced as oppressors of the people. *The right to use water should inhere in the land to be irrigated, and water rights should go with land titles.*"

The report was an immediate success; the first edition of 1800 copies was exhausted and 5000 more were printed. It brought Powell's appointment as a member of the Public Land Commission, laid out tasks which were taken up by the Geological Survey, and brought storms of condemnation from those who wanted to turn arid intermountain basins into humid-country farms. But dry seasons proved that Powell was right, and by 1888 he was authorized to establish an irrigation survey as one department of his larger bureau. Its function was to determine the extent of arid lands that could be irrigated, to select sites for dams and reservoirs, but not to construct or operate irrigation works. These, like mines and lumbering operations, were left to private enterprise.

This seems like a very moderate step, yet it led to serious trouble. Though little arid land could be irrigated, that little meant settlements, cultivation, fences and assignment of water rights. Each of these was anathema to the cattle barons, whose

Canyon's Conqueror

wealth came from free exploitation of national grazing land. In fury they went to legislators with demands that studies of irrigation be stopped, that Powell be disposed of, and that the Geological Survey itself be taught not to interfere with special privilege.

There already had been one attack upon Powell, inspired by members of old surveys who were not appointed to the new one. Though defeated, this opposition lingered and was ready to join critics who came to the aid of oppressed cattle kings. Their combined onslaught began in 1891, while Powell's maimed right arm was causing constant pain. As it throbbed his temper grew short, making work with hostile legislators almost impossible. Arguing gruffly for both geology and irrigation, he found the Survey's appropriation reduced from $719,400 to $631,940. A year later his enemies fully triumphed, first delaying the Survey bill until August and then whittling the appropriation to $379,885. Additional funds brought the total to $430,000, but irrigation was eliminated, many salaries were discontinued, and so much money was assigned to topographic mapping that geology was cut to the bone. Powell had to order a halt in field work, write letters warning of dismissal, and attempt to salvage results achieved instead of adding new ones. Scientists who refused to give up were carried on part pay or without salaries.

It was the familiar scheme of starving a bureau to drive its head out of office—a ruse only selfish men can resist. Powell soon found that his enemies were so strong and so bitter that opposition to them was hopeless. He already had talked of giving up directorship of the Survey; though reluctant to do so under pressure, he at last consented to resign because of "painful disability." This decision was rewarded by an increased appropriation for 1893–94. When that year closed he went to Baltimore for an operation on his arm, after which he took up full-time, vigorous work as director of the Bureau of American Ethnology.

Although Powell stands high among earth scientists of the

world, he wrote few technical papers and only two important books. One of these was the *Exploration of the Colorado River of the West,* a volume which combined rousing narratives of exploration with discussions of geology and transcribed Indian legends. The other was a more formal though smaller *Report on the Geology of the Eastern Portion of the Uinta Mountains,* published as a product of the second division of the United States Geological and Geographical Survey of Territories—the Hayden Survey—with Powell as geologist in charge.

Both reports show us a man whose outstanding concern was with the earth's fast-changing surface. To him formations and kinds of rock were minutiae; he described the former merely to use them and cared so little about the latter that he gave schists at the bottom of the Grand Canyon their still popular misnomer of granite. As for fossils—they were valuable for correlation and deserved careful study, but Powell never found them more than casually attractive.

Such an attitude must have seemed strange to men like James Hall, yet it worked to the vast advantage of earth science in America. In Powell's day few geologists on this side of the Atlantic realized that land changes were worth serious study or that land forms themselves were records of the past as significant as series of strata. But he was keen to detect the meaning of plateaus, escarpments and valleys, to see why some streams were more aged than others, to note the grit with which water had worn precipitous canyons and had scoured their one-time falls into rapids. While shooting rapids of the Green River, he realized that the stream must have been established before the Uintas arose, its muddy current cutting through bed after bed as the dome of strata heaved upward.

> The river [he wrote] had the right of way. In other words, it was running ere the mountains were formed; not before the rocks, of which the mountains are composed, were deposited, but before the formations were folded, so as to make a mountain range. . . . The emergence of the fold above the general surface of the country was little or no faster

Canyon's Conqueror

than the general progress of the corrasion of the channel. . . . The summit of the fold slowly emerged, until the lower beds of sandstone were lifted to the altitude at first occupied by the upper beds, and if these upper beds had not been carried away they would now be found more than twenty-four thousand feet above the river.

This was a dramatic statement of a vital idea; one that had grown through decades as an explanation for mountain gorges of Europe, India and the Appalachians, as well as for canyons in the West. It also gave strong support to the ideas of Hutton, Playfair and Lyell, which were still on trial in America and not too popular in Europe. Powell's proof that the Uintas rose no faster than the Green could erode its way through them ruled out King's sudden upheaval with its peaks five to eight miles high. "Thus aided," wrote one critic, "the gentle doctrine of geologic peace on earth gained a vast backward extension into periods of the past that had long been conceived as ages of violence."

The Green River had cut across rising, arched strata, but the Colorado was still wearing its canyons into a gently tilted plateau bounded on the west by faults. Working swiftly, the stream eroded bed after bed, leaving cliffs in resistant formations but making slopes in others that were weak. At some levels it exposed old flood plains and sand dunes; at others it revealed quiet sea floors or shoals where sea worms and trilobites once fed. With such matters Powell felt little concern, but he rejoiced to find records of buried mountain ranges deep down in the Grand Canyon. One could be traced in schists of the dark Inner Gorge—crumpled roots of peaks that were worn to a plain during the Archeozoic Era. In Keweenawan times the plain was a sea bed, receiving deposits of sediment which later tipped, broke and were heaved into long, narrow ranges like those of Nevada today. They, too, were reduced by erosion, though some of their cores still survive as ridges projecting upward into strata of Cambrian age. Powell published a poor diagram of them in his *Exploration of the*

The Story of the Great Geologists

Colorado River, but a good one in the later volume on the eastern Uintas.

Ancient ranges destroyed by erosion meant that mountains themselves were transient features of islands and continents. Powell told how they were being destroyed; since the processes of wasting were rapid it followed that rugged highlands were not very old. "Mountains cannot long remain as mountains," he wrote in 1876; "they are ephemeral topographic forms. Geologically all existing mountains are recent, the ancient mountains are gone." Before a critic who maintained that Eastern ranges were exceptions he shook his empty right sleeve and declared, "If the Adirondacks had been uplifted in Cambrian time they would have been worn down *over and over* AGAIN!"

He was right; the Appalachians have been worn down at least twice, and those that now form wooded ridges are barely 60,000,000 years old. Both they and younger ranges also owe their forms to erosion rather than to the uplift which gave them their height. For rain, streams and ice shaped ridges or peaks, just as grit-laden streams shaped the walls of Southwestern canyons. "The mountains were not thrust up as peaks, but a great block was slowly lifted, and from this the mountains were carved by the clouds—patient artists, who take what time may be necessary for their work. We speak of mountains forming clouds at their tops; the clouds have formed the mountains." He then showed that mountains (or hills, if the original blocks were not high) were bound to be reduced to lowlands of rolling near-level surface unless new uplift interfered by offsetting the work of streams.

Lowlands, great blocks—these words furnish keys to two radically new ideas. Before Powell no geologist had made it clear that erosion by rains and streams could do much more than make valleys between mountain ranges or hills. Nor had men working in Europe and the eastern United States been able to view uplift as more than arching or wrinkling produced as belts of land somehow shrank. Now erosion was found to

Canyon's Conqueror

wear highlands down to broad, featureless plains which in time might be reduced almost to sea level. The last stages of this process might be very slow; unless helped by solution, removal of the last few inches of a low plain "would require more time than all the thousands of feet that might have been above it." This detail, which seemed trivial in 1876, would one day play a large part in forming a new, intelligible concept of geologic ages.

As for uplift—crumpling remained important in regions such as the Alps or the Blue Ridge, but it could not explain plateaus hundreds of miles in width, where strata were almost horizontal. These must have been forced directly upward, with long fractures (faults) along boundaries between rising and stationary regions. Such faults were prominent west of the Grand Canyon, as well as in other parts of Arizona, Colorado and Utah. Gaps of one to two thousand feet were common, rocks on one side of each break having risen almost vertically.

Somewhat different was the Great Basin, where faulting had broken the land into wedge-shaped blocks which were miles or scores of miles in length. Some of these blocks tilted upward; others stayed where they were or even sank. The result of these movements was a series of elongate mountain ranges whose strata dipped into valleys floored with detritus worn from the elevated, inclined blocks. Other workers would apply this concept to the Wasatch Range, the Tetons and the Sierra Nevada, a tilted block four hundred miles long and one to three miles in height. Owens Valley would be recognized as a down-faulted segment, or graben, which broke and sank two miles or more after the Sierras were lifted.

In putting these ideas on paper, Powell followed rules of his own. Others might search through magazines and books, listing them in bibliographies, and giving full credit to authors whose ideas antedated their own. The major dealt with what he had seen, seldom referred to other authors, and preferred vivid descriptions to a scholarly style. He wrote of coal mines as "pots of pickled sunbeams," used emphatic reiterations, and

The Story of the Great Geologists

so disliked the ending "-al" that under him the Geological Survey (so termed by law) had geologic and paleontologic branches, and published a geologic atlas of the United States. Indeed, Powell and subordinates who followed him almost eliminated the term geologic*al* from all except official use.

Powell wrote his composite diary of boat trips down the Green and the Colorado in present tense and first person plural, so that readers run to boats and jump aboard, are swept out of them by waves, and are rescued by companions who somehow have got ashore. Geologic processes, however, were described in past tense. Yet even they fired Powell to rhetorical flights, so that after describing a cycle of erosion, deposition and renewed uplift he closed with this florid peroration: "Then again the restless sea retired, and the golden, purple and black hosts of heaven made missiles of their own misty bodies—balls of hail, flakes of snow, and drops of rain—and when the storm of war came the new rocks fled to the sea."

Twice Powell abandoned geology: once when he took charge of the ethnologic bureau and finally when he returned to it after being forced out of the Geological Survey. Each time he found in the study of primitive men a task for which he was qualified by temperament even more than by learning. To him it was natural that there should be standards beside those of civilization, for civilized standards would not fit primitive needs and knowledge. "When I stand before the sacred fire of an Indian village," he once wrote, "and listen to the red man's philosophy no anger stirs my blood." Nor was he vengeful toward those who in error had killed the men who started overland after Powell's own outbursts in 1869. A mistake admitted both explained and settled the matter. "That night I slept in peace, although the murderers of my men, and their friends, the *U-in-ka-rets,* were sleeping not five hundred yards away. When we were gone to the cañon, the pack-train and supplies, enough to make an Indian rich beyond his wildest dreams, were all left in their charge and all were safe; not even a lump of sugar was pilfered by the children."

Powell's first ethnologic work was published in 1874, to be followed by a long series of studies on language, mythology, tribal government, progressive culture and human evolution. Greatest of all was the *Indian Linguistic Families,* a monograph toward which Powell labored for years although it was published by a colleague in 1891. It divided red men of North America into fifty-eight basic groups, or stocks, marked by languages which differed as much as French and Arabic. A map showed the primitive home of each stock, with wide areas held by powerful, aggressive peoples and narrow ones occupied by those who were weak and of sedentary habit. Lack of intergrades and overlapping showed that early Indians were far less nomadic than many authorities had assumed.

From language to thought was not a great step, and Powell found time to deal with philosophy. It was the science of opinions, he found, and its progress reflected the knowledge on which those opinions were based. "The unknown known is the philosophy of savagery; the known unknown is the philosophy of civilization." In other words, man has progressed from observed facts which he explained but did not understand to a system of problems beyond certain knowledge, which he still tries to make meaningful. On the way he has often detoured into metaphysics, whose fatal error has been the "assumption that the great truths (or 'major propositions') were already known . . . and that by the proper use of the logical machine all minor truths could be discovered and all errors eliminated." Since both halves of this assumption were false its results could be nothing more than an orderly system of illusions, a "phantasmagoria."

A man who held such ideas was almost sure to attempt an organization of knowledge that would satisfy his own standards. Powell began this task about 1894, after leaving the Geological Survey. By 1898 he was ready to publish the first part of his system under the title of *Truth and Error; or, the*

The Story of the Great Geologists

Science of Intellection. It was followed by a series of articles to be reprinted with additions as a work on *Good and Evil.* A third part was unfinished when the author died at his Maine summer home in September 1902.

Philosophers have not praised Powell's system; not many scientists have read it. But every geologist looks upon him as a sort of American Murchison, who encouraged young men when they needed help, built up an adequate federal survey, and helped to make the nation's political capital a great capital of science. In the words of an associate, "He was to me not so much one of the common figures of daily life, as one of Plutarch's men. . . . Sincere he was, and truthful to the point of being unable to bring himself to hint the thing which is not, nor even to allow the shadow of deceit in his ways. Such sincerity existing in his own heart, begat a confidence in others which did not always meet its just return. . . . He was a generous man, kind to others and helpful; a combative and a brave, and always a self-contained man, who found in himself counsel sufficient for his need."

Enough, one might add, for two bureaus and the multitude of less assured men who came to him for help.

CHAPTER XX

Earth Blisters and Changing Land

EVERY LAND FORM has its meaning; those six words contain the kernel of Powell's geologic thought. In canyons, on plateaus and among bare mountains he worked out the origin of each form as it appeared in the West. His studies in the South and East were much less thorough, nor did he reduce processes to an order matching that devised for forms. Such work he left to other men, including one who became Powell's chief adviser and assistant on the Geological Survey. More than any other American, this man also linked the epoch of sweeping pioneer exploration with that of the detailed, precise and specialized researches which distinguish geology today.

Grove Karl Gilbert was one of those people who seem fitted by heredity for a science which deals largely with observations and demands a vivid memory. At the age of five or six he discerned wispy onion seedlings which his father had not noticed, insisting upon his discovery until his exasperated parent knelt and found the minute sprouts. Some years later, while skating on a bay, he examined ice cracks with such detail that his recollection of them proved useful after a half century. At the age of twelve he and a friend built a flat-bottomed boat so light that it could be held at arm's length and so speedy that owners of commercially built canoes refused to race against it. More significant, young Karl found out how to drive the boat forward at speeds of some yards per minute by bobbing up and down in the stern. This trick stuck in his mind for decades, until he was able to explain it in adult terms. "The motions I caused directly were strictly reciprocal," he decided, "the departures from initial positions being equaled by the returns. The indirect result of translation was connected with reactions between the water and the oblique surfaces of the boat."

These things happened in and near Rochester, New York, where Gilbert was born in 1843. His father had been something of a rolling stone; a fluctuating intellectual who had studied medicine, taught school, and finally settled down as self-taught portrait painter. In this field he worked with stubborn industry which was rewarded by honorary membership in the National Academy of Design. Yet poor earnings branded him dilatory if not foolish and shiftless among solid Presbyterian neighbors who could not see that painting was work. Still more caustic comments were made when the elder Gilberts withdrew from the church, to bring up their seven children without sermons or Sunday school. Such nonsense, such heathenish conduct! As if youngsters could grow up properly without knowing the threats of hell-fire or learning to fear their God!

Despite this lack Karl was a well-behaved boy who got along with other children and inspired a teacher to report: "His deportment has been unexceptionable, and he has been a most faithful, industrious and attentive student, meeting in every way my highest approbation." At fifteen he was a thin, narrow-chested youth whose chief trials were ill health and lack of money. The former kept him out of doors; the latter forbade many social amusements and compelled him to wear secondhand clothes whose cut and colors were not to his taste. Yet poverty was so far mitigated as to permit a classical course at the University of Rochester. It provided eight units

Earth Blisters and Changing Land

of mathematics, seven of Greek and six of Latin, but only one of geology. This seems to have been a dry, factual course under Professor Henry A. Ward, whose methods led Gilbert to suggest that science teachers would do well to "dwell on the philosophy of the science rather than its material." That recommendation, in different words, is still being made today.

Gilbert was nineteen when he graduated in 1862 as a tall, awkward, long-necked lad whose chin seemed much too large for his face. He lacked strength for service in the Army; having debts to pay, he tried to meet them by teaching at Jackson, Michigan. But he was too easygoing to manage a roomful of country youths whose ages were not much less than his own and whose vigor was much greater. Gilbert gave up when two thirds through his first year, going home to Rochester and "demoralizing" lack of employment. "I recall," he wrote his son almost fifty years later, "that I had no heart to do the various things that I supposed I very much wanted when I was too busy to find time. Waiting for something to turn up seems to be an occupation in itself, and anyone who can really utilize the time while he waits is to be congratulated."

Not that Gilbert had to wait very long, for an assistant was needed in Ward's Cosmos Hall. This was a remarkable combination of museum, taxidermist's shop and mail-order store which the professor had built up to supplement his university income. In it he sold stuffed and preserved animals, as well as skeletons, rocks, minerals, fossils and a miscellany of other things. His customers ranged from museums and colleges to private collectors, all served so well that Cosmos Hall still exists as Ward's Natural Science Establishment.

Gilbert began by copying labels but soon progressed to preparation of specimens and other responsible work. During 1867–68 he restored and mounted the skeleton of an Ice Age mastodon found near Cohoes Falls and still exhibited in the New York State Museum. While at Albany he studied the gorge of the Mohawk River, preparing a technical account for Hall's report on the mastodon and a popular article which was published in Moore's *Rural New Yorker*. The former explained that the elephantine beast had been mired in a huge and almost circular pit, or pothole, worn by rocks whirled round and round in a stream that was fed by a melting glacier. At least three hundred and fifty similar though smaller potholes had been worn upstream from Cohoes Falls by boulders in the modern Mohawk. Gilbert counted growth rings in gnarled cedars of the gorge, and by measuring their exposed roots concluded that thirty-five thousand years was "a minimum for the time . . . elapsed since Cohoes Falls were opposite the mastodon pothole."

Holes and receding falls proved so interesting that Gilbert resolved to devote his life to geology. Learning that a new survey of Ohio was being planned, he applied for a position, was refused, and became a volunteer assistant at fifty dollars per month for expenses. Next year (1870) he received a promotion and a salary, despite a rule which said that all appointments must go to Ohioans. During summers he traced glacial melting in the Maumee Valley and learned how streams had established their courses over newly formed moraines. In the winter he drew fossil plants and fish "in a style that has not been surpassed in this country"—to quote from an official report. Since this drawing had to be done in New York, Gilbert was able to attend theaters and lectures, as well as to meet scientists of the metropolis and near-by Yale.

In 1871 Lieutenant Wheeler organized his Geographical Surveys West of the One Hundredth Meridian, with Gilbert as the first and later as the chief geologist. By early May the expedition was afield in northeastern Nevada—a cumbersome affair of forty men with 165 horses and mules for which pasturage was a problem. After the party was divided into two and sometimes three smaller outfits, Gilbert spent eight months in almost constant movement which took him across desert basins and ranges to California, back to Arizona's high plateaus, and southwestward again to Yuma. The next season was spent in Utah and northernmost Arizona, while a third—

CLARENCE KING at the height of his official career.

Courtesy U.S. Geological Survey

JOHN WESLEY POWELL about the time he became director of the U.S. Geological Survey.

Ferdinand Hayden
and one of his field part
in camp at Red Butt
Wyoming, in August 187

Courtesy U.S. Geological Sur

Earth Blisters and Changing Land

Gilbert's last with the Wheeler Survey—extended from eastern Arizona into New Mexico. The trip out and back was made by stage from Pueblo, Colorado.

Those were days when the West was both new and wild; when towns were mere mining camps or stage stations where whisky, poker and billiards were the only recreations. At Arizona City Gilbert tried billiards but gave them up to write a column for the local paper. "The Free Press office," he noted, "is about 14 ft. square and includes the bed as well as the table and desk of the editor and all hands. Boxes serve as chairs & bottles as candlesticks. No stove. A dirt floor." Comments upon the editor himself were even less laudatory.

Near Prescott came word that a stage had been attacked; three members of the Survey were killed, with the driver and two other passengers. Near Marble Canyon Gilbert visited a ranch with a one-room house built half underground and two wagons to serve as bedrooms. Better conditions were found farther north where a Gentile, or non-Mormon, storekeeper welcomed Gilbert to "the most comfortable house I have seen for many weeks. It has four rooms and a housekeeper. The table cloth is white. The butter is good and the milk is cream. This is a combination of luxuries. . . ." Then came a review of recent sleeping quarters:

At Zion we furnished our own blankets & slept on the floor. At Rockville the same except we were furnished pillows. At Mt. Carmel we were given extra blankets & the lee side of a corn stack. At Toquerville I slept in a wagon box with the boy, at Workman's Ranch on the ground with the boy again. At Kanab in a bed on a bedstead alone, at Allendale ditto with the boy. At Circleville, ditto, ditto.

From desert basins and dry, high plateaus Gilbert went to spend the winter in humid Washington. There he and two other geologists rented a room not far from the White House and Wheeler's office but boarded two miles away to make sure of exercise. On Sundays they took long walks among farms east of the Anacostia or through forests along the Potomac.

The Story of the Great Geologists

In the evenings they played cards, took dancing lessons, and studied German, the last without much success. Playing catch near the market went much better until police broke up the game in deference to pedestrians and windows of near-by stores.

In January of 1874 Gilbert went to a dance at Powell's and there met Fanny Porter. They took Sunday walks, went on drives, saw Barnum's museum, and were married on November 10 at the bride's home in Cambridge. Till summer their home was a rooming house, but with late autumn the Gilberts and their month-old daughter settled in a Washington suburb that was favored by geologists. Thereafter the father's journal abounded in such entries as "Today we begin 3 pts. cream and 3½ qts. milk," which led to "Bessie 13 lbs.," and was followed by "Rattle, 0.25." With a literary flourish, housekeeping costs were charged to "Dame Durden" or, when in haste, to "D.D."

Gilbert had met Powell in 1872 and the two men began to share results of their explorations. When military control became too repressive, Gilbert first quit the Wheeler Survey and then, three weeks after his marriage, began work under Powell. He helped illustrate the latter's great Uinta report, devising novel diagrams which showed both modern scenery and ancient geologic structures. Gilbert also revisited plateaus of Arizona and Utah and then began the first study which brought him lasting fame.

The Henry Mountains lie in southeastern Utah, a cluster of domes and eroded peaks which rise a mile above the surrounding plateau, to heights exceeding eleven thousand feet. Gilbert examined them hastily in August 1875 and from their porphyritic crests looked down upon strata that dipped with the slopes and then became level beneath the surrounding desert. Did such structure mean "volcanic cones of elevation" which had arched beds into domes as they pushed toward the surface?

The question was answered in 1876, when Gilbert's party spent two months among the Henry Mountains. While topog-

Earth Blisters and Changing Land

raphers prepared a map the geologist climbed peak after peak and traced seven thousand feet of strata into the rolling plateau. In this series were two thick formations of shale, one three thousand feet above the other, and each containing biscuit- or dome-shaped intrusions which once were doughy magma. In that state they had welled up through narrow

JUKES BUTTE IN THE HENRY MOUNTAINS

Gilbert's figure retouched to show intrusive rocks (stippled) standing in cliffs above arched strata which are underlain by another laccolith.

necks or fissures, spreading out when they reached the shale bands. There they baked level subjacent strata while arching those above. Though this may have happened rapidly, the weight of a vast overburden would not let the arched strata break.

Gilbert called these intrusions laccolites; "literally, lakes of stone." He also explained that they were not superficial structures; that they formed thousands of feet underground and were exposed by erosion which wore bed after bed away until the intrusions remained as heights upon the denuded plateau. This process, Gilbert wrote, "has given the utmost variety of exposure to the laccolites. In one place are seen only arching strata; in another, arching strata crossed by a few dikes; in another, arching strata filled by a net-work of dikes and sheets. Elsewhere a portion of the laccolite itself is bared, or one side is removed so as to exhibit a natural section. Here the sedimentary cover has all been removed, and the laccolite stands free, with its original form; there the hard trachyte itself has been attacked by the elements and its form is changed. Somewhere, perhaps, the laccolite has been destroyed and only a dike remains to mark the fissure through which it was injected."

On one point Gilbert was very clear; laccolites—we now spell the word *laccoliths*—were not deeply buried versions of volcanoes or lava flows which dot dry Southwestern plateaus. Their rocks were utterly distinct, and so was their distribution. Thus one series of 118 volcanoes showed no trace of subterranean arching, while laccoliths of the Henry group bore no hint of cones or flows. They were bodies of a special kind, whose shape, position and even size obeyed rules quite different from those of eruption.

Gilbert left the Henry Mountains in November; by March of the following year his report upon them was written. Printed copies were dated 1877, but difficulties in reproduction of plates delayed publication more than a year. Meanwhile Powell undertook his report on arid regions, and Gilbert began to gauge streams and examine irrigation projects. In Salt Lake City he called upon Brigham Young, finding the aged leader in excellent humor after a hearty meal. "We had a nice talk," Gilbert reported. "There were a dozen present, half of whom were Church Dignitaries, and I found in a few minutes that I was talking only nominally to Brigham, but really to his advisers. They talked to the point, and appreciated what I was at, but he strayed as badly as Dr. Hayden."

That last clause may explain why another man than Hayden became head of the new United States Geological Survey in 1879. Under King, Gilbert was appointed geologist with a salary of $4000 and an office in Salt Lake City, where he took charge of work in the Great Basin. Its ancient lakes were his

Earth Blisters and Changing Land

chief concern until 1881, when Powell abolished King's regional divisions and called Gilbert back to Washington.

Mormon settlers had long been familiar with steep banks and level terraces that followed the Wasatch front and encircled lesser mountain ranges in the region of Great Salt Lake. Gilbert had examined them briefly in 1872, concluding that they were made by waves, and outlining a vanished body

"WAYS AND MEANS" Gilbert's illustration under the heading "How to Reach the Henry Mountains" in his official report.

of water which once covered twenty thousand square miles. Later he named it Lake Bonneville, to honor a military explorer of the early West.

The task now was to reconstruct this lake, tracing its story in detail. Gilbert found that it had begun during the Ice Age, after millions of arid years during which the mountains were deeply eroded and detritus was piled against their slopes. Then came a long, moderately humid epoch which sent rivers into connected intermontane basins from which there was no escape. The water, therefore, formed a lake—a lake that became nine hundred feet deep, received dissolved salts from streams, and lost water only by evaporation. Its shore deposits are now unknown, but ninety feet of clay that covered its bottom now lie under later beds.

Then came another arid epoch, in which the first Lake Bonneville vanished and its basin was scoured by winds. But the climate again grew damp as glaciers formed on peaks and rains fell on intermediate slopes. Their water revived the lake, which deepened until its waves beat on cliffs a thousand feet above the basin. They cut steep banks and built terraces, fashioned spits in shallow water, piled bars across the heads of bays. At last the lake overflowed to the northward, swiftly cutting a gorge down to resistant limestone. There the water level stood for many years, till returning aridity reduced it to lower and lower levels. At last only Great Salt Lake and a few small ones remained as shrunken, shallow and saline remnants of the once great Bonneville.

Gilbert completed his map of Lake Bonneville late in 1880; next year he returned to Washington, intending to write his report. But the Survey was growing faster than Powell could guide it, and the would-be author found himself adviser, confidant and responsible assistant to his overburdened chief. Gilbert designed maps, planned researches, consulted with field men; he advised that manuscripts be filed, returned or published, and rewrote at least one large paper while his own was put aside. Advance chapters on Lake Bonneville did appear, but 1890 drew to a close before the whole work was published as a quarto monograph. The author was quite prepared for reviews which praised it while calling some chapters prolix, but chuckled when papers treated it as news.

I have been interrupted [he remarked], by a reporter. He interviewed me today on Lake Bonneville and came in this evening with his report for me to revise. He says that he has sold it to the N. Y. Tribune and it will probably be telegraphed tonight to appear in tomorrow's (Thursday's) paper. It strikes me as very comic that what I found out years ago should be sent to N. Y. by telegraph instead of mailing the MS. But the reporter never heard of it before, nor have the readers of the Tribune.

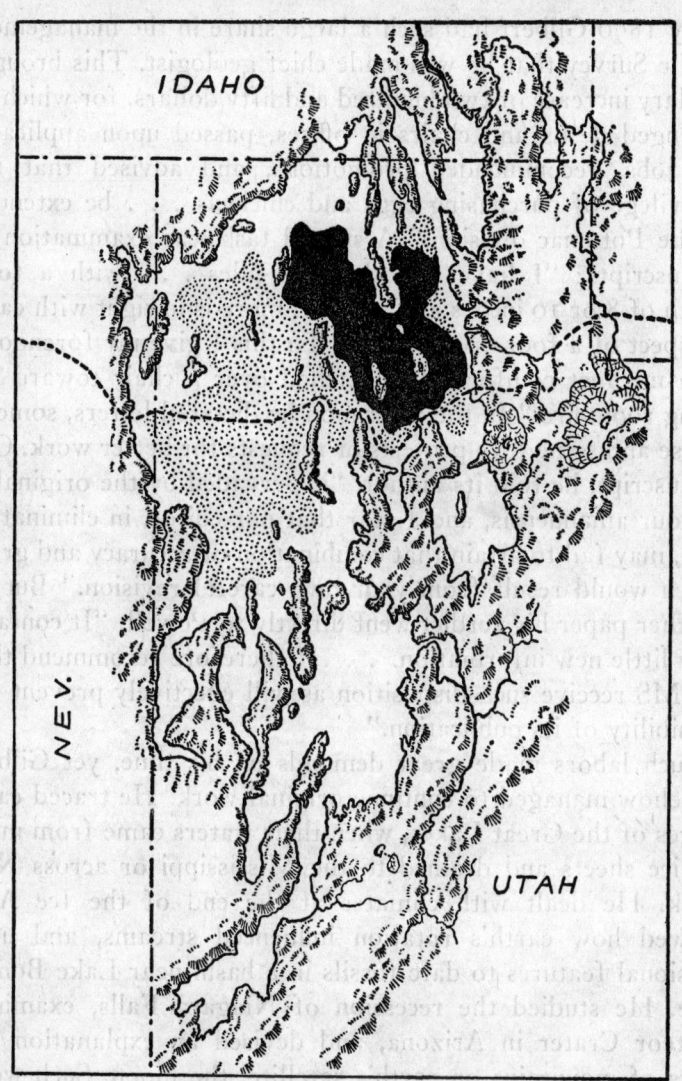

LAKE BONNEVILLE

As mapped by Gilbert. Black marks Great Salt Lake; stippled areas are desert; the letter G marks glaciers in the Wasatch Mountains. The modern U.S. Highway 40 has been added as a broken line.

The Story of the Great Geologists

By 1890 Gilbert had such a large share in the management of the Survey that he was made chief geologist. This brought a salary increase of two hundred and fifty dollars, for which he arranged desks and chairs in offices, passed upon applicants for jobs, recommended promotions, and advised that the "privilege of purchasing eggs and chickens . . . be extended to the Potomac division." A special task was examination of manuscripts: "I have already read piles . . . with a total depth of 8 or 10 inches and as much more is in sight with early prospect of a foot or two more. To this I give my forenoons, with my feet comfortably stretched over a chair toward my living room fire." In the afternoon he dictated letters, some in praise and some with protests or requests for better work. One manuscript, he told its author, "is pervaded by the originality of your amanuensis, and I fear that our editor, in eliminating that, may fail to attain that combination of accuracy and grace which would result from your own careful revision." But on another paper his verdict went directly to Powell. "It contains very little new information. . . . I therefore recommend that the MS receive such disposition as will effectively prevent the possibility of its publication."

Such labors made great demands on his time, yet Gilbert somehow managed to continue original work. He traced early stages of the Great Lakes, when their waters came from melting ice sheets and drained to the Mississippi or across New York. He dealt with climates at the end of the Ice Age, showed how earth's rotation influenced streams, and used erosional features to date fossils in a basin near Lake Bonneville. He studied the recession of Niagara Falls, examined Meteor Crater in Arizona, and devised an explanation for rings of mountains on earth's satellite, the moon. Such work aroused the ire of one congressman, who raged: "So useless has the Survey become that one of its most distinguished members has no better way to employ his time than to sit up all night gaping at the moon."

This was in 1892, when cattle legislators almost wrecked

Earth Blisters and Changing Land

the Survey and forced Powell to resign. Gilbert already had refused the directorship as well as a position at Cornell, and he probably welcomed demotion from the rank of chief geologist. Under the new regime he spent three summers on the plains of Colorado, reviewed his work on the Great Lakes, and dealt with complex problems of balance between continents and ocean basins. In 1899 he visited Alaska—not at government expense, but as one of two dozen scientists who were guests of E. H. Harriman. Though the trip lasted only two months, Gilbert was able to study some forty ice streams, describing the preglacial surface and the erosion of deep coastal valleys, or fiords. In one of these ice seemed to plow along the bottom for two thousand feet beyond its apparent front at the surface of the sea.

This record suggests an undisturbed life, yet Gilbert's was full of worry. His daughter died in 1883; his wife became a helpless invalid; his two sons needed companionship which field work often prevented. With a salary which seldom exceeded four thousand dollars, Gilbert had to earn additional money by lecturing, writing articles, and revising a school textbook on physical geography. In good years these efforts piled up savings of seven to nine hundred dollars, but house repairs combined with illness might leave less than a hundred. The year 1893 was prosperous, for although Gilbert's salary had been reduced he received one thousand dollars for twelve lectures at Columbia University. Still, money was not to be wasted, and Gilbert took evenings in New York to mend clothes. "My overcoat sleeve lining misbehaved and I set it right. Then while I was about it I inserted various other stitches in various garments, each time saving 800% profit."

Mrs. Gilbert died in 1899; two years later the widower resumed field work in the Great Basin. This is the region which lies between the Sierras, the Wasatch, and Arizona's plateaus, with an eastern extension toward the Great Plains in New Mexico and Texas. Powell had characterized it as "desert valleys between naked ridges"; Gilbert himself had described it briefly in reports of the Wheeler Survey. There he suggested that the basin region had been broken into a series of north-south blocks along almost vertical faults that appeared late in Jurassic times. Uplift of some blocks matched depression of others; both were connected with vulcanism; both were accompanied by erosion which reduced upthrown blocks to ridges and peaks and provided loose fragments that were washed from high sectors to low ones. The result was a series of jagged mountains and almost level valleys where the streams that flowed after sudden rains quickly disappeared. Many of those basins had been sites of lakes, some older and some of about the same age as Lake Bonneville.

Traveling through Utah in 1901, Gilbert found this explanation too simple. Tilted ranges revealed ancient folds or were crossed by faults which in turn were cut by breaks of a later age. Both showed that the region had been crumpled during late Jurassic times, when arched ranges were being filled with granite in what now is the Sierra Nevada. Then began millions on millions of years of erosion during which folded mountains were worn into rolling prairies. At last came the events of Gilbert's first theory: great breaks that were roughly parallel, dividing the country into blocks that either heaved upward or sank. These changes, indeed, were still going on in faults that made barren cliffs near the Wasatch and on several lesser ranges.

Some men might have glossed over the contrast between old theory and new one; Gilbert announced happily, "Among my interesting finds are a number of mistakes made by Gilbert, one of the Wheeler geologists." Later statements might have elaborated on those discoveries had not maps made in 1901 been destroyed by accident. Their loss so disturbed Gilbert that he published almost nothing, avoiding work on the Basin Ranges for a dozen years. Even then he gave them only two weeks, of which "the last two days . . . were profitable." In 1916 he found a fault at the eastern edge of the Cascades and then examined block mountains near Klamath Lake—blocks

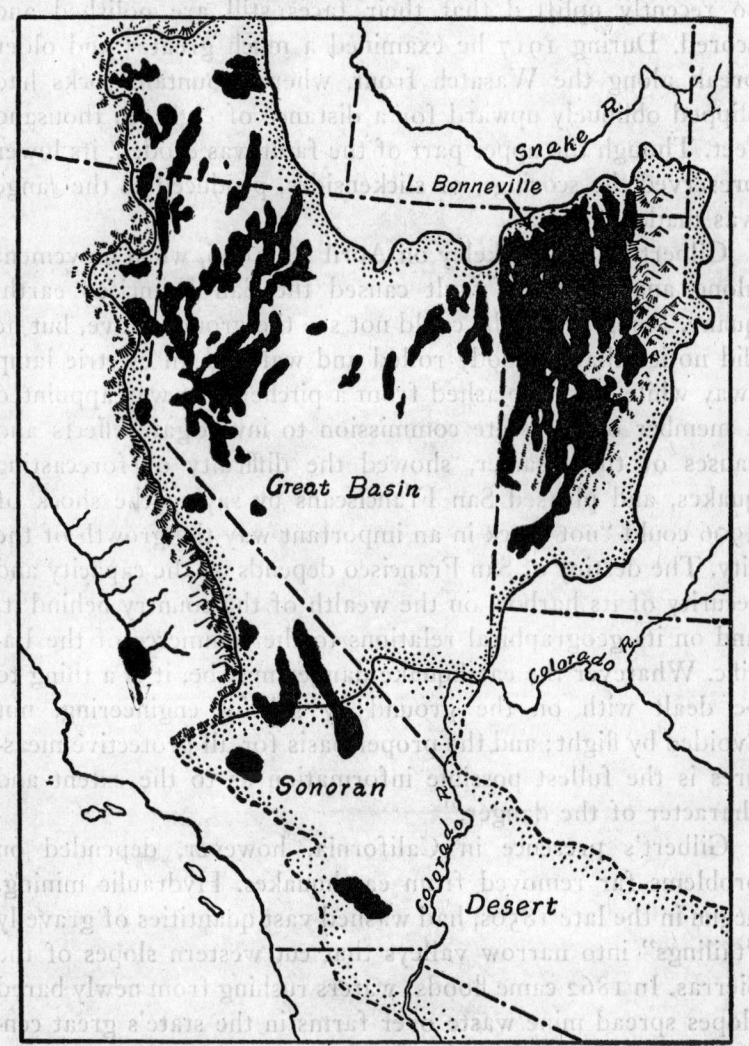

BONNEVILLE AND OTHER ICE AGE LAKES

This map follows Gilbert by extending the Great Basin northward into Oregon. Modern usage stops it a few miles north of Nevada and California. Lakes are shown in black.

so recently uplifted that their faces still are polished and scored. During 1917 he examined a much greater and older break along the Wasatch front, where mountain rocks had slipped obliquely upward for a distance of eighteen thousand feet. Though the upper part of the fault was eroded, its lower preserved the scorings, or slickensides, produced as the range was made.

Gilbert was in Berkeley on April 18, 1906, when movement along another great fault caused the San Francisco earthquake. Being in bed, he could not see the ground move, but he did notice that his body rolled and watched an electric lamp sway while water splashed from a pitcher. He was appointed a member of the state commission to investigate effects and causes of the disaster, showed the difficulty of forecasting quakes, and pleased San Franciscans by saying the shock of 1906 could "not check in an important way the growth of the city. The destiny of San Francisco depends on the capacity and security of its harbor, on the wealth of the country behind it, and on its geographical relations to the commerce of the Pacific. Whatever the earthquake danger may be, it is a thing to be dealt with on the ground by skillful engineering, not avoided by flight; and the proper basis for all protective measures is the fullest possible information as to the extent and character of the danger."

Gilbert's presence in California, however, depended on problems far removed from earthquakes. Hydraulic mining, begun in the late 1850s, had washed vast quantities of gravelly "tailings" into narrow valleys that cut western slopes of the Sierras. In 1862 came floods; waters rushing from newly bared slopes spread mine waste over farms in the state's great central valley. This released a flood of protests and lawsuits, to be followed by injunctions and a commission so restrictive that miners at last petitioned for a study by the United States Geological Survey. It began a thorough investigation, with Gilbert in charge of both field and laboratory work on the transportation of debris.

Earth Blisters and Changing Land

Field work took most of four years; then a hydraulic laboratory was set up at the State University in Berkeley. Till 1910, when illness overtook him, Gilbert had active charge and lived at the Faculty Club. The results of his laboratory work appeared in the *Transportation of Débris by Running Water,* a remarkably thorough examination of the ways in which streams move rock rubbish and of the influence of such factors as volume, speed and slope. It called for extensive computation, and probably seemed useless to miners who wanted to wash more gravel for gold without causing undue harm. Their needs were met by a second report, on *Hydraulic-Mining Débris in the Sierra Nevada,* a study in river engineering that is one of Gilbert's most brilliant works. In it he traced tailings from uplands to mountain canyons, thence to piedmont farm lands, and on through rivers of the Great Valley till they reached San Francisco Bay, where they built thirteen thousand acres of shoals into tidal lands. These reduced the amount of water sweeping in and out of the Golden Gate, causing a tidal bar to shift shoreward about one thousand feet. Reclamation of marshes had a similar effect, and in time might cause both shoreward shifting and troublesome decrease in depth. Limit reclamation, Gilbert warned, or face the prospect of damage to harbors on San Francisco Bay.

Gilbert found the "working up of the tidal-prism problem in the bays mighty interesting," even though he started out with an erroneous theory. The work called for a return to California in 1914, with studies of tidal ebb and flow at the Golden Gate. By midwinter he was reading proof, a task that stretched through three full sets as he found error after error. "Think," he exclaimed in May of 1915, "think of letting 'a phenomena' pass me several times in the ms. and twice in proof!" Unwilling to let such errors go, he revised so carefully that the book was not published until 1917.

Gilbert's illness of 1909–10 was grave: a "brain trouble of apoplectic character" with warning of complete collapse. He might, perhaps, have worked on as Owen did; instead he took almost constant rest and followed a rigid schedule laid down by his doctor. In consequence he began to improve and by summer of 1910 did some work for the report on debris. A year later he went to Massachusetts, where he wrote a great deal, studied ripple marks, and tried salt-water canoeing. But one trip among breakers was almost too hard; it "set the wheels going in my head," he reported, "and I didn't amount to much for a day or two."

By 1912 he must have seemed almost normal: a tall, slightly stooped man with full beard and gray hair who paddled on the Potomac, read aloud, and played billiards in tournaments. He learned to drive an electric automobile—his sister's—but gave up gasoline cars in Washington because of nervous strain. Thereafter he said little about illness, except when excessive heat in the Sierra foothills stopped his work in 1914, sending him to the cool comfort of San Francisco Bay.

Still, the warning of collapse remained—a warning that turned to reality in the spring of 1918. Gilbert was then almost seventy-five, still an active member of the Survey, and en route to California with a partly written report on the *Structure of the Basin Ranges*. He expected to spend about a hundred and fifty dollars of government money on field work—"as the journey will be at my own expense the draft on Survey funds will be small." But illness struck as he stopped to visit his sister, and after a painful week he asked his son to come and clear up some business matters. Twelve days later he played one last rubber of cribbage, and on May 1 he died.

A list of Gilbert's publications might seem scattered, purposeless, with subjects ranging from ancient ice sheets to blisters of molten rock. Yet in all we should find one recurrent theme: that the land changes in rational ways and with rational results. Thus an ancient stream whirled rocks, eroding great potholes, and the Mohawk River did likewise on a less gigantic scale. A glacier spread moraines in the Maumee Valley; streams crossed them by flowing from low place to low

Earth Blisters and Changing Land

place, just as they did on moraines in other glaciated regions. Laccoliths of the Henry Mountains were eroded, not on a hit-or-miss basis but according to principles that were valid in ranges near Lake Bonneville. The laws for transportation of debris apply to streams in the East or Africa as well as to those that carry tailings from Sierran slopes.

In short, what Gilbert did was to build up a subscience, physiography, which deals with the earth's surface and processes by which it is being changed. It was not Gilbert's alone, of course, for other Americans made their contributions, as did geologists of Europe. But Gilbert knew little of European work, while conditions in an area of broad ice sheets, bare mountains and great arid wastes gave his contributions a peculiarly native flavor. To Powell's basic grouping of arid-country land forms they added a structure of process and cause which extends far beyond lands of little rain yet maintains their clear-cut character.

CHAPTER XXI

Glaciers to Galaxies

BELOIT, WISCONSIN, a century ago was a farm-land village whose rutted streets stopped at fields or turned into roads that wandered across moraines. Flowers bloomed on pastured prairies in the spring; in the fall great flocks of passenger pigeons alighted on blackened acres from which farmers had burned the stubble after reaping wheat. On winter nights wolves came from near-by woods to bicker with dogs whose noisy courage was strengthened by picket fences and sheds. These quarrels went on hour after hour unless blasts from muzzle-loading muskets drove the wild participants away.

Toward this settlement, in 1846, creaked two loaded prairie schooners owned by John Chamberlin. The wagons paused awhile in town, for the Chamberlins needed supplies after camping along the road from Illinois. Then whips cracked and wheels rolled northwestward to a grove where the family established themselves in a log cabin, also with its picket fence. The barrier kept a three-year-old toddler from straying over prairies or into woods.

The little boy, christened Thomas Chrowder, had been born in September of 1843 on the crest of a terminal moraine near Mattoon, Illinois. When four years old he listened, deeply

Courtesy U.S. Geological Survey

GROVE KARL GILBERT
one of America's most brilliant geologists.

Courtesy Rollin T. Chamberlin

THOMAS CHROWDER CHAMBERLIN
at his office desk.

Glaciers to Galaxies

impressed, while his father described the laying of a cornerstone for the new college in Beloit and explained what the ceremony meant. When eight, while his brothers boasted of grown-up adventures to come, young Thomas showed his bent by declaring, *"I'm* going to school till I can teach the very best school in the state!" When the other boys jeered Father came to his aid. "Boy, don't you back out on that. Stick to it!"

From their cabin the Chamberlins moved to a farm a mile or more away. Tom helped quarry blocks of limestone to build a foundation for the new house and shoveled sand for mortar. With his brothers he puzzled over the long, straight "snakes"—fossil mollusks—in the buff Ordovician strata. How had the reptiles wriggled into beds of stone, and what made them petrify? Clay balls in the sand pit were much less puzzling, for the Chamberlins held the prevailing opinion that these were growing boulders, or "niggerheads," not yet old enough to be large and hard.

John Chamberlin was both farmer and preacher; a circuit rider who believed in education and led long religious arguments which deeply impressed his four sons. Growing up, they attended Beloit College, a Congregational school which offered a classical course modeled on that of Yale. Tom was stirred by philosophy and mathematics, but his approach to geology was that of a hostile skeptic. Its ages conflicted with Holy Writ, as did its account of earth's beginning from a ball of incandescent gas. Young Chamberlin was willing to investigate, yet sure that he could answer such falsehoods with scriptural evidence. To his surprise the "errors" proved convincing—convincing and so attractive that he wanted more of the subject. His enthusiasm was encouraged when the professor anticipated his query about grades by asking, "Well, Chamberlin, would you like to know who stood *next?*"

The new convert to science graduated in 1866, while Gilbert was writing labels at Ward's after his failure as a teacher. But Chamberlin was neither shy nor unwell; he was a husky young fellow more than six feet tall, with wide shoulders, a commanding presence and a record in college football. With these went profound respect for fact, a belief that the "greatest genius is probably a genius for seeing realities." This cropped up when he took his first examination for a teacher's certificate and encountered the question "If the third of 6 is 3, what would the fourth of 20 be?" Outraged, the candidate answered: "The fourth of 20 is 5 under any and all circumstances and is not affected by any erroneous supposition that may be made in respect to a third of 6."

Now came seven years of alternate teaching and study as principal of a public high school, as professor of sciences in a state normal school, and as a graduate student at the University of Michigan. In the high school Chamberlin conducted laboratory work, led field trips and gave popular lectures—all novel additions to education in the 1860s. At the normal school he taught so inspiringly that several of his students decided to follow science professionally. One of these was Rollin D. Salisbury, a brilliant but timid and pessimistic youth who finished the four-year course in three years, taught, and then entered Beloit College, where Chamberlin had become professor. Salisbury's choice was geology, though his first college teaching included zoology and botany. In time he became Chamberlin's close friend and aide, as well as his collaborator in field work, teaching and authorship of textbooks.

Wisconsin established a geological survey in 1873, eleven years after abolishing one headed by James Hall. Chamberlin apparently laid plans for the new organization and prompted a friendly legislator who saw to it that plans became law. Friends expected him to be made state geologist and were indignant when that job went to another man, with Chamberlin as a mere assistant assigned to the south central part of the state. His district had no mines, no very ancient rocks, no great exposures of fossiliferous strata. How could he do himself justice in a region of such slight interest?

These complaints diminished in 1875, when politics displaced the chief geologist, replacing him with a physician who

Glaciers to Galaxies

held the position only one year. In February of 1876 Chamberlin took charge and for three years conducted the Survey without interference.

These were years of intense labor and commensurate achievement, for beside directing a scientific staff of seven, Chamberlin carried on field work throughout the eastern part of the state. There he studied water resources and soils, investigated shell marls and clays, and proved to doubting Americans that the microscope could be used to study such unconsolidated deposits. He concluded that mounds of dolomite near Racine were old coral reefs and that beds of granular rock about them once were limy sands worn as waves battered and broke dead corals. In an examination of the lead and zinc district he corrected Owen on minor points, but supported both Owen and Hall by concluding that ores lay too near the surface for deep mining to be worth while.

Most important of all, however, were his studies of glacial deposits, or drift. Geologists had written of *the* Glacial Epoch; of a drift sheet left by one great glacier which spread like a viscous fluid over high lands and low. Chamberlin proved that Wisconsin's drift sheets and ridges dated from at least two glacial epochs separated by an interval during which ice "withdrew entirely from our territory, if not from the Canadian highland." He also found that both ice sheets had spread in irregular lobes, some so distinct from one another that they appeared to be separate glaciers. So far as the Middle West was concerned, these lobes showed most plainly during the second frigid epoch, when eight of them advanced toward the so-called Driftless Area of the upper Mississippi Valley.

Survey field work closed in 1879, but Chamberlin took two more years to write and publish reports. For nine years he had been professor at Beloit College; he now reduced this connection to a lectureship and joined the United States Geological Survey under Major Powell. Heading a new Glacial Division, he was directed to investigate the "terminal moraine that enters the U.S. on the north border of the Territory of Dakota and stretches thence southward and eastward in sinuous course probably to the Atlantic." He first prepared a now classic paper on the conditions necessary for artesian wells and then, with Salisbury as his assistant, examined the Driftless Area. There, "divided by highlands, led away by valleys, consumed by wastage where weak, self-perpetuated where strong," ancient glaciers had repeatedly closed around a hilly region but only once had ridden across it. The report contrasted glaciated with non-glaciated landscapes, described products of weathering, and weighed theories to account for deposits named for the Alsatian town of Loess. In 1900 Chamberlin would return to the problem, suggesting that the loess of North America and Europe had begun as ground-up rock spread over flood plains by streams that flowed from melting ice sheets. When waters receded and the flood plains dried, winds whirled the rock flour away in dust clouds which settled on near-by hills.

In the midst of these studies Chamberlin was approached by regents of the University of Wisconsin. That institution, they said, needed a new type of president: a man of science, vision and vigor who would turn their pseudo-Eastern college into a real university. Chamberlin answered that geology was his proper field, but the regents were insistent. The task needed doing and he could do it. Did he not owe this service to education, as well as to his state?

Chamberlin at last agreed to the offer, took a year to complete work for the Survey, and assumed office at Madison in 1887. There he met some opposition, especially from faculty members who held that university presidents should be clergymen. How could a geologist envisage students' spiritual needs? And who could be secured to deliver baccalaureate sermons?

The geologist answered by giving them himself, making a deep impression on professors, students and public. He reassured those who feared science by strengthening work in the classics and arts; he justified the regents' faith by reforming a cumbrous organization, regrouping and improving elective courses, and placed emphasis on graduate instruction as well as faculty research. Under him the university offered its first

Glaciers to Galaxies

agricultural short course, set up the Midwest's first psychological laboratory, and instituted an extension program to serve all parts of the state. Chamberlin also brought new men into engineering departments and planned a school in which

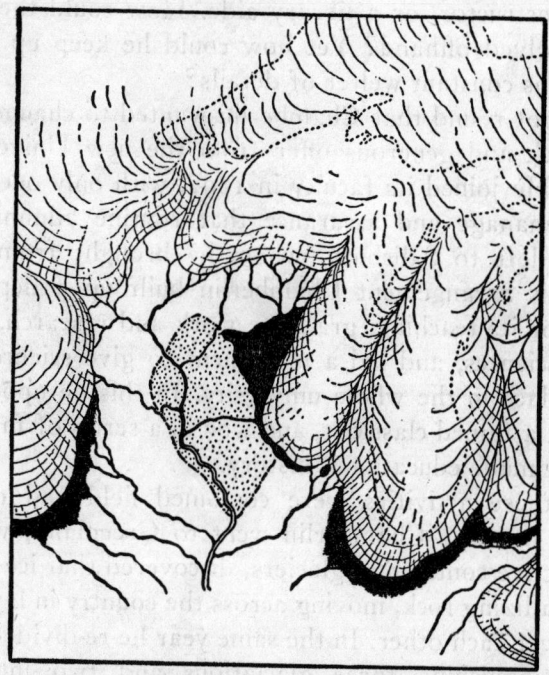

LOBES OF TWO CONTINENTAL ICE SHEETS

Chamberlin's map showing lobes of ice converging about the so-called Driftless Area (stippled). The extent of the marginal lakes (black) was uncertain.

history, political science and economics would collaborate. Instead of merely telling what had happened, they would train students to analyze and participate in affairs.

Here was a life's work accomplished in four years; accomplished without fuss or confusion and with full legislative support. But the university had also to be run, and day-to-day administration was a task that wore Chamberlin down. For he was first and always a scientist; a man who could overlook no details, act on no inadequate data, render no casual judgments. He delegated few tasks and gave minor matters the same careful thought that preceded major decisions. For a minor matter might really be vital—vital, for instance, to an erring student, a young instructor, or a library aide. How could their problems be solved offhand? Yet how could he keep up creative work in this constant welter of details?

Word got round that Chamberlin wanted to change, bringing a quick and generous offer from the new University of Chicago. He joined its faculty in 1892, with only one department to manage and assurance that routine administration might be left to Salisbury, whom he brought from Beloit. Under this arrangement Chamberlin built up a department famous for its teaching, graduate work and research, became dean of sciences, and led a movement to give science an important place in the whole university. In this he opposed the president, a famed classicist, and won in a series of faculty debates that made educational history.

With these activities were combined field and office researches. In 1894 Chamberlin went to Greenland, where he studied active continental glaciers, discovered that ice behaved like crystallizing rock, moving across the country in layers that slipped over each other. In the same year he re-divided the Ice Age, distinguishing three glaciations and two interglacial epochs. In 1896 he increased their numbers to four and three; the same divisions recognized today, though modern names differ a bit from those proposed by Chamberlin. Changes came from attempted refinements which, though erroneous, he did not oppose.

NAMES OF GLACIAL EPOCHS

Chamberlin, 1896	Modern Usage	European Epochs
4. Wisconsin	4. Wisconsin	4. Würm
3. Illinoian	3. Illinoian	3. Riss
2. Iowan	2. Kansan	2. Mindel
1. Kansan	1. Nebraskan	1. Günz

Glaciers to Galaxies

The Nebraskan of modern usage is merely Chamberlin's Kansan renamed, the latter term being transferred to the epoch which he called Iowan. European epochs are listed to show their agreement with those of North America.

From glacial deposits Chamberlin turned to the cause of glacial periods. Like most geologists of his time, he had been trained to accept Laplace's theory that the earth once was a ball of hot gas which reached far out in space. In time, however, the ball condensed, forming a white-hot liquid sphere whose material cooled by radiation till its surface began to harden. With continued cooling the crust became thicker and stronger. At last it was so strong that the molten core could break through only in occasional eruptions from fissures or volcanoes.

When erupting lavas reached the surface they encountered the atmosphere in an almost primeval state. It was thicker by far than that of today; a heavy gaseous blanket rich in water vapor. It also contained fifty times the present amount of carbon dioxide, or enough to provide all the carbon now locked up in coal, petroleum, limestone and black shale. Since both carbon dioxide and water vapor are absorbers of heat, the thick atmosphere made a more than tropical cover for the solidifying earth.

Ages passed with their inevitable changes. As the planet cooled water condensed, to gather in lakes, rivers and seas. Carbon dioxide combined with weathering rocks or was taken from the air by early plants whose remains colored strata, formed beds of coal, or turned into petroleum. The atmosphere became less thick and less moist, which meant that it absorbed less heat from the sun and lost what it did absorb more and more rapidly. The result was a progressive cooling of climates, which went on until both polar regions grew frigid while longer and colder winters descended upon temperate regions. In a time when these winters grew specially cold the Pleistocene ice age gripped our planet with its expanding glaciers.

The Story of the Great Geologists

This, Chamberlin admitted, was a dramatic and compelling picture with its "trend of a moribund earth toward a cold senility." The thickened crust already was cold; the sun, which still gave warmth, was cooling; the earth body was shrinking and cracking, drinking in water and absorbing gas at an inordinate rate. The planet must continue to cool off and dry up, both glaciers and deserts would keep on growing, and a final winter followed by desiccation seemed inevitable. The airless, waterless moon foretold earth's future, with the Ice Age as a mild forewarning of stringent times to come.

The theory was simple, spectacular, familiar, yet it met difficulties when recurrent glacial epochs were discovered, with intervening epochs of warmth. Still more serious was evidence of long-term fluctuations; of alternating warmth and coolness, humidity and dryness throughout geologic history. Long before this evidence included proofs of repeated glaciations during Pre-Cambrian and later eras, Chamberlin had rejected the Laplacian theory of progressively changing climates and had proposed a substitute.

This new theory began with two facts already known and accepted: that carbon dioxide absorbs heat and that great fluctuations in its amount must cause climatic changes. One such change had to follow epochs of uplift, for rocks raised into mountains or plateaus would take up carbon dioxide in the process of weathering. This would lower temperatures; not over the land alone, but above and in the oceans. Ocean waters then would absorb more carbon dioxide, again reducing the atmospheric blanket and depressing temperatures. Changes thus became cumulative, until at last they went so far that winter's snows could not be melted by the sunshine of shortened summers. Instead, fields of snow packed into ice which moved down mountains or spread across lowlands with increasing thickness and speed. With it, of course, went arctic weather, which brought more snow to the margins of ice sheets and thereby helped them spread.

A cold period with its glaciers was produced; now to melt

the ice away by climates that must moderate. Here Chamberlin relied on erosion, which followed every great uplift and reduced high lands to low. As it did so it released carbon dioxide—not directly, but through a calcareous compound that lost its gaseous portion before settling in the sea. There it formed the great limestone formations that characterize every period of low land, while carbon dioxide entered the air and began to treasure heat. Long before the last mountains were reduced to base level, a warming blanket of air was turning the last glaciers into muddy streams.

We summarize this theory bluntly; Chamberlin advanced it cautiously in 1897. Critics frowned their doubts but he kept on working, and nine years later supported his first conclusions with a study which showed that marine circulation of cold epochs must differ from that of warm. During the latter, said Chamberlin, evaporation was so rapid in equatorial regions that surface waters became very salty. This meant that they grew heavy and sank, flowing poleward to warm ocean abysses and spread their moderating influence over the entire globe. They also reduced the sea's absorption of carbon dioxide, since warm water holds less of this gas than does cold.

Then came uplift with its cooling, initiated in the air. Evaporation soon was reduced, and in time the poleward flow of warm water ceased. In its place came heavy, cold water which sank in frigid zones and flowed to the tropics, chilling the ocean depths as it went. The result was still more rigorous climates, with diversity controlled by features of the land as well as by surface currents.

Uplift, weathering, carbon dioxide, seas; they combined in a theory which seemed to explain alternations of climate throughout geologic time. Since the Timiskaming, if not the Keewatin, earth's history has been a sequence of long periods of erosion with increasingly warm or equable climates, followed by shorter ages of uplift with cooling and sometimes glaciation. Four of these early ice ages have been traced, each coinciding with high lands and even with desert conditions.

The Story of the Great Geologists

The late Tertiary was another time of elevation, followed by Pleistocene glaciations and very rapid erosion. If Chamberlin was right, we are living in an intermediate epoch which has enough warmth to melt remaining ice streams and ice sheets, but too little to halt the flow of cold water equatorward from the poles. Therefore the earth clings to its glacial climatic pattern, with deserts, humid lowlands and snow fields in identical latitudes. These contrasts are made more extreme by the fact that lands still are high.

If Chamberlin was right—that no one can tell, for geologists have made few tests of his theory. Some thought it too speculative for attention; some complained that it could not account for repeated glacial and interglacial epochs of the Pleistocene. Chamberlin thought it might, but was so engrossed in a greater problem that he could not make sure. This problem was the origin of our earth, and with it the solar system. For if the Laplacian theory could neither explain earth's climatic changes nor agree with them, could it account for the origin of our planet? If it failed to do so, what theory was better—or must a new one be framed?

These questions were posed in 1897, when Chamberlin published his first paper on glacial climates and the British lion of science, Lord Kelvin, took geologists to task for dealing too freely with time. Kelvin already had announced that the sun was a progressively cooling ball whose heat was produced by contraction. Its age therefore was set by physical laws which also put a limit upon planets, including the earth. King had used these laws to get his figure of 24,000,000 years, published in 1893. Kelvin allowed more latitude, for his computations gave the earth a minimum age of 20,000,000 years and a maximum of 40,000,000. But these extremes were not open to question. Within them geologists must compress all time from Recent to earliest Keewatin, plus the formative eras.

Chamberlin read Kelvin's published address with amazement, finding serious geologic errors and unjustified physical assumptions. Two years later he tore the restrictions to shreds

Glaciers to Galaxies

while the more conservative Britishers gasped at New World temerity.

The field was reopened; now to explore it. Chamberlin's first step was to secure the aid of a brilliant young astronomer named Moulton, with whom he applied exacting tests to Laplace's theory itself. Defects had appeared and more soon were found; attempts to repair them failed and uncovered new difficulties. By 1900 both Moulton and Chamberlin were ready to abandon the time-honored theory as hopeless, as fatally in conflict with facts. Equally grave faults were found in rival ideas, including those that tried to build our planet from formless swarms of meteorites. If the two Chicagoans desired an acceptable explanation of earth's origin, they plainly must devise their own.

This they did during the years 1900–04, with assistance from a chemist and a mathematical physicist. Their first efforts, however, were futile: long weeks of computation which proved only that one nebula was unlikely to collide with another. Eruptions of gas from the sun proved more promising; they led to studies of tidal disruption and new computations to appraise the chance of another sun having passed close to ours. These were followed by studies of orbits, rotation, momentum, heat; of remote spiral nebulae and planetary atmospheres. Toward the end of 1904 Chamberlin rechecked data, reviewed conclusions, and summarized both in a paper which proposed a new account of earth's origin called the planetesimal hypothesis. It was published by the Carnegie Institution, which had helped to finance the work.

Chamberlin's hypothesis began with the sun, a bit larger than it is today, but with the same belts of gas storms and eruptions above and below its equator. For eons it traveled through space, shooting out clouds of incandescent gas that rose many thousands of miles but generally sank back to its surface. Though far from quiescent, it produced nothing that might turn into planets.

At last another star came near—near in an astronomical

The Story of the Great Geologists

sense, which probably means approach within many millions of miles. This star was smaller than the sun, yet its gravitative pull set up great tides that drew out or released great bolts of gas that dwarfed all ordinary eruptions. They swung toward the star as it passed; then escaped its control and traveled in orbits around their parent sun. They looked, thought Chamberlin, much like the coiled arms of spiral nebulae.

Though the bolts were gaseous when they left the sun, their material soon condensed into solid bodies termed planetesimals. Since they moved in varied orbits and at different speeds, the planetesimals were bound to collide. In thick parts of the bolts these collisions soon built up spheres which would become planets or their satellites.

At this stage the earth appeared as a smallish sphere close to the sun, though much farther from it than the spheres which would be Venus and Mercury. It continued to collide with planetesimals and grow, most often by minute accretions, but sometimes by capturing masses thousands of times larger than our biggest meteorites. Reaching about half its present diameter, the young planet began to hold air whose gases came in with planetesimals or emerged during early volcanic eruptions. At first the atmosphere was thin and dry, since water molecules were so light that they readily escaped the earth's modest gravitative control. But as size and mass increased more vapor was held; vapor that at last condensed and began to fall as rain. From that time onward our world was what it is now: a rigid, rocky spheroid surrounded by air and partly covered by water. When life finally appeared it did so among surroundings which differed in no essential from those encountered today.

The planetesimal hypothesis was superior to others, yet it had to be improved. Chamberlin had used spiral nebulae as a pattern for his primitive solar system, constructing a restoration which showed great coils of incandescent gas revolving around the sun, with thick clouds ready to condense into planetary cores. But spiral nebulae proved to be galaxies: whole "island universes" of stars as large as the one to which our sun belongs and much like it in shape. Chamberlin therefore gave them up as models, made new computations, and designed a more realistic arrangement of ancestral gas bolts and knots. These later studies indicated that the outer group of large planets, which are not very dense, came from bolts shot toward the second star. The inner group of compact planets, including Mercury, Venus, the earth and Mars, were descended from tidal bolts that shot in opposite directions and so did not go very far. These two groups could be arranged in pairs beginning with Neptune-Mars and ending with Jupiter-Mercury. If Pluto, which was unknown to Chamberlin, possessed an opposite it has not been identified.

Another radical change was made when comets and most meteorites were traced back to eruptions independent of the small passing star. Even today, many of the sun's prominences shoot out at speeds almost great enough to keep them from drifting back. In ancient times, said Chamberlin, there must have been many eruptions which went so far that they were drawn by other bodies in our galaxy and so began to travel in great ellipses. Today they swing through space or fall upon the earth, along with bodies which really are remnants of the once abundant planetesimals. Although some are massive, the vast majority are no larger than a grain of wheat.

These refinements were made during twenty-eight years; years in which Chamberlin accomplished many other tasks. In 1904–06 he and Salisbury published a massive three-volume textbook of geology, the most thorough ever written in English. It restated the planetesimal hypothesis, proposed new causes for vulcanism, depression and uplift, defined new geologic periods, and used cyclic changes in the earth as a basis for major time divisions. No more vital and influential general work has appeared since Lyell upset catastrophism with the first edition of his *Principles*.

In 1909 Chamberlin joined a commission to report on needs for philanthropic work in China, under Rockefeller auspices.

The Story of the Great Geologists

With his son (also a geologist) as assistant, the now aging professor crossed the Pacific and traveled through Asia for five months, visiting cities and peasant villages, measuring Pre-Cambrian glacial deposits, and studying family farms. He was deeply impressed by Chinese industry, thrift and common sense, and saw great opportunities for technical education. But the effects of disease impressed him still more, and he agreed with other members of the commission when they recommended medical improvement as the first and most helpful step. The result was Peking Union Medical College, in which young Chinese could be trained to care for their countrymen.

The investigation was strenuous. By day Chamberlin consulted officials, visited schools, examined farms and traveled in boat, cart or sedan chair. At night he sketched railroad routes, outlined scientific changes, noted possibilities for development of natural resources. In Szechwan he walked most of four hundred miles to study terracing, planting and intensive cultivation that yielded three to five crops per year. Reaching Mukden, he fell ill and welcomed the long train trip to Moscow as a chance for rest. An expedition to Turkestan was canceled, but with his son he zigzagged across Europe from Norway to Rumania.

Such effort was too much for an ailing man of almost sixty-six. It brought to an end Chamberlin's field work and marked further reduction of his activities at the university. For forty-three years he had labored in office, classroom and field, seldom taking a vacation and teaching elementary classes until 1906. Now he limited himself to a single advanced course in the principles and theories of earth sciences, leaving simpler subjects to young men and the brilliant but now dogmatic and caustic Salisbury. Students who liked to contrast these two leaders told of a farmer who found Chamberlin digging in drift, requested an explanation, and was given a stirring account of the Ice Age and its varied moraines. "Thanks, mister," said the farmer as Chamberlin closed. "Yesterday I tried to find out some of those things from a fellow with big eyes and little, pointed whiskers. But, d'you know what the cuss did? He just glared at me and snapped, 'This-is-a-Wisconsin-moraine-don't-ask-questions-I've-got-work-to-do!' "

Chamberlin had work to do, too: a simple book on the *Origin of the Earth* and a series of technical articles dealing with *Diastrophism and the Formative Processes*. In them he concluded that an earth built up from planetesimals was much more than a solid ball. It was a complex of materials and forces imbued with resources for combination, adjustment and compression which might—even must—raise continents, depress marine basins, and build range after range of mountains as it shrank in circumference. The articles appeared between 1913 and 1921, with writing sometimes interrupted by illness or carried on when the author could barely read his own large script. Yet on good days he might dash off fifty to sixty pages or with magnifying glass pore for hours over volume after volume in search of useful data.

As the series drew to a close, Chamberlin saw that new work must be done on the planetesimal hypothesis. Other men were producing alternative explanations; alternatives often derived from his own ideas, but ingenuously preserving faults of the Laplacian theory. Chamberlin criticized these efforts in reviews and then began a two-volume revision of the *Origin of the Earth* which would both bring the book up to date and apply results of that improvement to the planet's development. The first volume, called *The Two Solar Families*, was published on September 25, 1928, and the first copy was hurried to Chamberlin's home for an afternoon celebration. He planned to take a month's rest and then begin the second volume, but within that month he fell gravely ill, and on November 15 he died.

Sources and References

This list omits a number of rare works to which adequate reference is made in the text. Additional publications of American geologists are recorded in the bibliographies of Nickles and Miss Thom.

General

Adams, F. D. *The Birth and Development of the Geological Sciences.* Baltimore, 1938. An authoritative work, but does not deal with Lyell and later geologists.

Agar, W. M., R. F. Flint and C. R. Longwell. *Geology from Original Sources.* New York, 1929. A book for college students.

Dana, E. S. (editor). *A Century of Science in America.* New Haven, 1918. Deals especially with the influence of the *American Journal of Science.*

Geikie, Archibald. *The Founders of Geology,* second edition. New York, 1905. Interesting and reliable.

Mather, K. F., and S. L. Mason. *A Source Book in Geology.* New York, 1939. Excellent collection of extracts from geologic classics.

Merrill, G. P. *Contributions to the History of American Geology.* U.S. National Museum, Annual Report for 1904, pp. 189–733. 1906. Supplanted by the same author's book of 1924.

——. *Contributions to a History of American State Geological and Natural History Surveys.* U.S. National Museum, Bulletin 109. 1920. Data on work, appropriations, publications, etc.

——. *The First One Hundred Years of American Geology.* New Haven, 1924. A large and very useful book based on the National Museum publication of 1906.

Nickles, J. M. *Geologic Literature on North America, 1785–1918.* U.S. Geological Survey Bulletins 746–47. 1923.

——. *Bibliography of North American Geology, 1919–28.* U.S. Geological Survey Bulletin 823. 1931.

Thom, E. M. *Bibliography of North American Geology, 1929–39.* U.S. Geological Survey Bulletin 937. 1944. An index to biographies and other material published since 1928.

Woodward, H. B. *History of the Geological Society of London.* London, 1907.

Sources and References

Zittel, K. A. *History of Geology and Palæontology.* Translated by M. M. Ogilvie-Gordon. London and New York, 1901. More comprehensive but less detailed than works by Geikie and Adams.

Chapter I

Herodotus. *Herodotus.* With an English translation by A. G. Godley. London and New York, 1921–22.

Lones, T. E. *Aristotle's Researches in Natural Science.* London, 1912.

Pliny. *The Natural History of Pliny.* Translated with notes and illustrations by J. Bostock and H. T. Riley. London and New York, 1890–1900.

Ross, W. D. (editor). *The Works of Aristotle,* Vol. 3, *Meteorologica.* Translated by E. W. Webster. Oxford, 1931.

Singer, C. J. *From Magic to Science.* London, 1928. Chapter I deals with "Science Under the Roman Empire."

Strabo. *The Geography of Strabo.* Translated by H. L. Jones. New York and London, 1924–31.

Tozer, H. F. *Selections from Strabo, with an Introduction on Strabo's Life and Works.* Oxford, 1893. Unusually interesting.

Wethered, H. N. *The Mind of the Ancient World. A Consideration of Pliny's Natural History.* London and New York, 1937. Revealing and stimulating.

Chapter II

Holmyard, E. J., and D. C. Mandeville (editors). *Avicennae de Congelatione et Conglutinatione Lapidum.* Paris, 1927.

Steele, Robert. *Roger Bacon and the State of Science in the 13th Century.* Oxford, 1921.

Steno, Nicolaus. *Nicolai Stenonis de Solido intra Solidum naturaliter contento dissertationis prodromus.* Facsimile edition. Berlin, 1904.

———. *The Prodromus of Nicolaus Steno's Dissertation Concerning a Solid Body Enclosed by Process of Nature within a Solid.* Translated by J. G. Winter. New York, 1916.

Chapter III

Condorcet, M. J. A. C. de. *Œuvres de Condorcet,* Vol. 3, pp. 317–47. Brunswick, 1804, or Paris, 1847–49. Sole source of biographical data on Guettard.

Cuvier, G. C. L. D. *Recueil des éloges historiques,* Vol. 2, pp. 339–74. Paris, 1819. Only biography of Desmarest, whose name is spelled "Desmarets" throughout.

Desmarest, Nicolas. *Géographie physique,* Vol. 1. Paris, 1794. Articles on Antrim, Auvergne, Basalte, Courans and Guettard.

———. *Mémoire sur l'origine et la nature du basalte à grandes colonnes polygones, determinées par l'histoire naturelle de cette pierre, observée en Auvergne.* Mem. Acad. Royale des Sciences for 1771, pp. 705–75. Paris, 1774.

Geikie, Archibald. *Geological Sketches at Home and Abroad.* New York, 1882. Excellent chapter on the volcanoes of central France, pp. 74–108.

Sources and References

Guettard, J. E. *Atlas et description minéralogiques de la France, entrepris par ordre du roi par MM. Guettard et Monnet.* Paris, 1780.

Chapter IV

Beck, Richard. *Abraham Gottlob Werner, Eine kritische Würdigung des Begrunders der modernen Geologie, zu seinem hundertjährigen Todestage.* Berlin, 1918.

Buch, Leopold von. *Geognostische Beobachtungen auf Reisen durch Deutschland und Italien.* Bonn, 1802 and 1809.

Frisch, S. G. *Lebensbeschreibung Abraham Gottlob Werners.* Leipzig, 1825.

Moro, A. L. *De Crostacei e degli altri marini Corpi che si truovano su Monti.* Venice, 1740. Translated as *Neue Untersuchung der Veränderungen des Erdbodens nach Anteitung der Spuren von Meerthieren und Meergewachsen,* etc. Leipzig, 1751 or 1775.

Werner, A. G. *Kleine Sammlung Mineralogischer Berg-und-Huttenmannischer Schriften.* Leipzig, 1811. Contains table of contents for proposed great work.

———. *Neue Theorie von der Entstehung der Gänge.* Freiberg, 1791. Introduction complains of unauthorized publications.

———. *Kurze Klassifikation und Beschreibung der verschiedenen Gebirgsarten.* Dresden, 1787. "Sent to press" in 1777, and printed by a friend.

Chapter V

Ferguson, Adam. *Minutes of the Life and Character of Joseph Black, M.D.* Transactions Royal Society of Edinburgh, Vol. 5, pp. 101–17. 1805.

Hutton, James. *Theory of the Earth.* Transactions Royal Society of Edinburgh, Vol. 1, pp. 209–304. 1788.

———. *Theory of the Earth, with Proofs and Illustrations.* Edinburgh, 1795.

Kay, John. *A Series of Original Portraits.* Edinburgh, 1838. Hutton in Vol. 1.

Kirwan, Richard. *Examination of the Supposed Igneous Origin of Stony Substances.* Transactions Royal Irish Academy, Vol. 5, pp. 51–81. 1793(?).

———. *Geological Essays.* London, 1799. "On the Huttonian Theory of the Earth," pp. 433–99.

Playfair, John. *Illustrations of the Huttonian Theory of the Earth.* Edinburgh, 1802. Reprinted in *Works of John Playfair, Esq.,* Vol. 1, 1822. Perhaps the first geologic work in English which is modern in style as well as substance.

———. *Biographical Account of the Late Dr. James Hutton, F.R.S. Edin.* Transactions Royal Society of Edinburgh, Vol. 5, pp. 39–99. 1805. Reprinted in *Works,* Vol. 4, 1822.

——— and E. Jeffrey. *Biographical Account,* in *Works of John Playfair, Esq.,* Vol. 1, pp. xi–lxxvi. Edinburgh, 1822.

Chapter VI

D'Aubuisson de Voisins, J. F. *Memoire sur les basaltes de la Saxe, accompagné d'observations sur l'origine des basaltes en general.* Paris, 1803. Translated as *Basalts of Saxony,* Edinburgh, 1814.

Sources and References

———. *Traité de géognosie.* Paris, 1828-35.
Jameson, Robert. *Elements of Geognosy.* Edinburgh, 1808.

Chapter VII

Eyles, V. A., and J. M. Eyles. "On Different Issues of the First Geological Map of England and Wales." *Annals of Science,* Vol. 3, pp. 190-212. 1938. Critical examination of changes in Smith's map.
Judd, J. W. "The Earliest Engraved Geological Maps of England and Wales," *Geological Magazine,* Dec. 4, Vol. 5, pp. 97-103. 1898.
———. "William Smith's Manuscript Maps," *Geological Magazine,* Dec. 4, Vol. 4, pp. 439-47. 1897.
Phillips, John. *Memoirs of William Smith, LL.D.* London, 1844. A rare work; important but rather colorless.
Sheppard, Thomas. *William Smith, His Maps and Memoirs.* Proceedings Yorkshire Geological Society, new series., Vol. 19, pp. 75-253. 1917.
Smith, William. *A Delineation of the Strata of England and Wales, with part of Scotland,* etc. 1815 or later printing.
———. *Memoir to the Map and Delineation of the Strata of England and Wales with part of Scotland.* London, 1815. This memoir and map are very rare. An exceptionally fine copy of the map, once owned by William Maclure, is preserved in the library of the Academy of Natural Sciences of Philadelphia.

Chapter VIII

Bonney, T. G. *Charles Lyell and Modern Geology.* New York, 1895.
Lyell, Charles. *Elements of Geology,* several editions. The first American edition, with hand-colored frontispiece, was published at Philadelphia in 1839.
———. *The Geological Evidences of the Antiquity of Man.* London and Philadelphia, 1863.
———. *Principles of Geology. Being an Inquiry into How Far the Former Changes of the Earth's Surface Are Referable to Causes Now in Operation.* Several editions.
———. *A Second Visit to the United States of North America.* New York and London, 1849; second edition, 1855.
———. *Travels in North America in the Years 1841-42.* London and New York, 1845. An important commentary on the United States and upon Lyell's own ideas.
Lyell, Mrs. *Life, Letters and Journals of Sir Charles Lyell, Bart.* London, 1881.

Chapter IX

Clark, J. W., and T. M. Hughes. *Life of the Rev. A. Sedgwick.* Cambridge, 1890. Detailed and mildly partisan.
Flett, J. S. *The First Hundred Years of the Geological Survey of Great Britain.* London, 1937.

Sources and References

Geikie, Archibald. *Life of Sir Roderick I. Murchison.* London, 1875.
Hudson, John. *A Complete Guide to the Lakes.* Kendal, 1853. Contains Sedgwick's account of geology of the Lake District.
Murchison, Roderick. *Siluria. The History of the Oldest Known Rocks Containing Organic Remains,* etc. London, 1854.
———. *The Silurian System.* London, 1839.
Sedgwick, Adam. *A Synopsis of the Classification of the British Palæozoic Rocks,* etc. London and Cambridge, 1855. Pages v–xcviii illustrate Sedgwick's attacks upon Murchison.

CHAPTER X

Agassiz, E. C. *Louis Agassiz, His Life and Correspondence.* Boston and New York, 1885.
———. *A Journey to Brazil.* Boston and New York, 1895.
Holder, C. F. *Louis Agassiz, His Life and Work.* New York, 1893.
Marcou, Jules. *Life, Letters and Works of Louis Agassiz.* New York, 1896.

CHAPTER XI

Fisher, G. P. *Life of B. Silliman.* New York, 1866.
Goode, G. B. *The Beginnings of American Science.* U.S. National Museum, Annual Report for 1897, Part 2, pp. 409–66. 1901.
Hall, C. R. *A Scientist in the Early Republic. Samuel Latham Mitchill, 1764–1831.* New York, 1934.
Maclure, William. *Observations on the Geology of the United States.* Philadelphia, 1809; revised edition, 1817.
Mitchill, S. L. "Observations on the Geology of North America," in Cuvier's *Essay on the Theory of the Earth,* pp. 321–428. New York, 1818.
Morton, S. G. "A Memoir of William Maclure," *American Journal of Science,* Vol. 47, pp. 1–17. 1844.
Volney, C. F. *A View of the Soil and Climate of the United States of America.* Translated by C. B. Brown. Philadelphia, 1804.

CHAPTER XII

Eaton, Amos. *A Geological and Agricultural Survey of the District Adjoining the Erie Canal.* Albany, 1824.
———. *A Geological and Agricultural Survey of Rensselaer County in the State of New-York, to which is annexed a Geological Profile.* Albany, 1822.
———. *Geological Textbook.* Albany, 1830; second edition, 1832.
———. *An Index to the Geology of the Northern States.* Albany, 1818; second edition, Troy, 1820.
McAllister, E. M. *Amos Eaton, Scientist and Educator.* Philadelphia, 1941. Probably the most detailed biography of an American geologist.
Marcou, Jules. "Biographical Notice of Ebenezer Emmons," *American Geologist,* Vol. 7, pp. 1–23. 1891.

Sources and References

Chapter XIII

Clarke, J. M. *James Hall of Albany, Geologist and Paleontologist.* Albany, 1921.

Hall, James. Volumes of the *Palæontology of New-York* and other publications listed in the bibliography by Nickles. Many must be examined, and with care, to reveal the scope of Hall's genius and the faults in his method of work.

Chapter XIV

Anonymous. "The Mineral Lands of the United States," *United States Magazine and Democratic Review,* Vol. 8, pp. 30-42. 1840. A typical review of Owen's 1839 report.

Hendrickson, W. B. *David Dale Owen: Pioneer Geologist of the Middle West.* Indiana Historical Collections, Vol. 24. 1944. An authoritative biography.

Owen, D. D. *Report of a Geological Exploration of Part of Iowa, Wisconsin and Illinois,* revised and illustrated. Senate Documents, 28th Cong., 1st session, 7, no. 407. 1844.

———. *Report of a Geological Reconnoisance of the Chippewa Land District,* etc. Senate Executive Documents, 30th Cong., 1st session, 7, no. 57. 1839.

———. *Report of a Geological Survey of Wisconsin, Iowa, and Minnesota; and Incidentally of a Portion of Nebraska Territory.* Philadelphia, 1852. The great "Owen Report"; an example of fine bookmaking and writing, as well as field research.

Owen, R. D. *Twenty-seven Years of Autobiography. Threading My Way.* New York, 1874. By David Dale Owen's brother.

Parker, N. H. *Iowa as It Is in 1856; a Gazetteer for Citizens, and a Handbook for Immigrants.* Chicago, 1856. Reproduces extracts from Owen's report of 1852.

Podmore, Frank. *Robert Owen. A Biography.* New York, 1907. Biography of David Dale Owen's father.

Chapter XV

Harrington, B. J. *Life of Sir William Logan.* Montreal, 1883.

Logan, W. E., and others. *Report on the Geology of Canada* ["*Geology of Canada*"]. Geological Survey of Canada, Report of Progress to 1863. Montreal, 1863; map, with James Hall as joint author, published in 1869.

Logan, W. E. Technical papers listed by Nickles.

Chapter XVI

Coleman, A. P. *Ice Ages, Recent and Ancient.* New York, 1926.

Dawson, J. W. *Fifty Years of Work in Canada.* Edited by Rankine Dawson. London and Edinburgh, 1901.

———. *Relics of Primeval Life.* Chicago, 1897. A convenient summary of Dawson's ideas on *Eozoon.*

———. "Review of the Evidence for the Animal Nature of Eozoon cana-

Sources and References

dense," *Geological Magazine,* Dec. 4, Vol. 2, pp. 443–49; 502–06, 545–50. 1895.

———. *Specimens of Eozoon canadense and Their Geological and Other Relations.* McGill University, Peter Redpath Museum, Notes on Specimens. Montreal, 1888.

Rice, W. N. (editor). *Problems of American Geology.* New Haven, 1915. Chapters on the Canadian Shield, pp. 43–80, 81–161. Rather old, but graphic.

CHAPTER XVII

Wilmarth, M. G. *The Geologic Time Classification of the United States Geological Survey Compared with Other Classifications.* U.S. Geological Survey, Bulletin 769. 1925.

CHAPTER XVIII

Cope, E. D. "F. V. Hayden," *American Geologist,* Vol. 1, pp. 110–13. 1888.

Emmons, S. F. *Biographical Memoir of Clarence King.* National Academy of Sciences, Biographical Memoirs, Vol. 6, pp. 25–55. 1907.

Hague, J. D. *The Mining Industry, with Geological Contributions by Clarence King.* U.S. Geological Exploration of the Fortieth Parallel, Vol. 3. 1870.

King, Clarence. "Active Glaciers within the United States," *Atlantic Monthly,* Vol. 27, pp. 371–77. 1871.

———. "The Age of the Earth," *American Journal of Science,* Ser. 3, Vol. 45, pp. 1–20. 1893. Reprinted in Smithsonian Institution, Annual Report for 1893, pp. 335–52. 1894.

———. *Systematic Geology.* U.S. Geological Exploration of the Fortieth Parallel, Vol. 1. 1878. Interesting for its remarkable illustrations as well as its content.

Powell, J. W. *Ferdinand Vandiveer Hayden.* U.S. Geological Survey, Ninth Annual Report, pp. 31–38. 1889.

Schmeckebier, L. F. *Catalogue and Index of the Publications of the Hayden, King, Powell and Wheeler Surveys.* U.S. Geological Survey, Bulletin 222. 1904.

White, C. A. *Memoir of Ferdinand Vandiveer Hayden, 1839–87.* National Academy of Sciences, Biographical Memoirs, Vol. 3, pp. 395–413. 1893.

CHAPTER XIX

Davis, W. M. *Biographical Memoir of John Wesley Powell.* National Academy of Sciences, Biographical Memoirs, Vol. 8, pp. 11–83. 1915. Principal source of information about Powell.

Gilbert, G. K. (editor). *John Wesley Powell; a Memorial to an American Explorer and Scholar.* Chicago, 1903. Consists of reprints from Vols. 16–17 of The Open Court.

Powell, J. W. *Exploration of the Colorado River of the West and Its Tributaries.* Smithsonian Institution, 1875.

———. *Report on the Geology of the Eastern Portion of the Uinta Moun-*

Sources and References

tains and a Region of Country Adjacent Thereto. U.S. Geological and Geographical Survey of Territories [Powell], 1876.

——. *Report on the Lands of the Arid Region of the United States, with a More Detailed Account of the Lands of Utah.* 45th Congress, 2nd session, House Executive Document 73, 1878; second edition, 1879.

Stanton, R. B. *Colorado River Controversies.* New York, 1932. A critical examination of Powell's record of his work along the Colorado River; deals also with the claims made for James White.

Summers, R. A. *Conquerors of the River.* New York, 1939. Semifictional in style; deals with Powell's first trip down the Colorado. Very interesting.

Chapter XX

Davis, W. M. *Biographical Memoir of Grove Karl Gilbert.* National Academy of Sciences, Memoirs, Vol. 21, 5th memoir, 1926. Detailed criticism of Gilbert's work, as well as a biography.

Gilbert, G. K. *Lake Bonneville.* U.S. Geological Survey, Monograph 1. 1890.

——. *Report on the Geology of the Henry Mountains.* U.S. Geographical and Geological Survey of the Rocky Mountain Region, 1870; second edition, 1880.

Chapter XXI

Chamberlin, R. T. *Biographical Memoir of Thomas Chrowder Chamberlin, 1843–1928.* National Academy of Sciences, Biographical Memoirs, Vol. 15, pp. 305–407. 1934. The principal source on T. C. Chamberlin; written by his son, also an eminent geologist.

——. *Memorial of Rollin D. Salisbury.* Geological Society of America, Bulletin, Vol. 42, pp. 126–38. 1931.

Chamberlin, T. C. "A Group of Hypotheses Bearing on Climatic Changes," *Journal of Geology,* Vol. 5, pp. 653–83. 1897.

——. *On a Possible Reversal of Deep-Sea Circulation and Its Influence on Geologic Climates.* Proceedings American Philosophical Society, Vol. 45, pp. 33–43. 1906.

——. *The Origin of the Earth.* Chicago, 1916.

——. *The Two Solar Families.* Chicago, 1928.

—— and R. D. Salisbury. *Geology.* New York, 1904–06; second edition, 1905–07. A monumental textbook.

Merrill, G. P. "The Development of the Glacial Hypothesis in America," *Popular Science Monthly,* Vol. 68, pp. 300–22. 1906.

Index

Numbers in heavy type refer to illustrations on plates, which face the pages indicated. Titles of books and periodicals are in italics.

Aar Glacier, 118, 119
Academy of Natural Sciences, 131
Agassiz, Alexander, 192
Agassiz, Jean Louis Rodolphe, 68, 111, 156, **95**
  childhood, 112
  death, 123
  education, 112
  Fossil Fishes, 117
  Glacial System, 119
  glaciers, 116
  Lake Superior region, 121
  lectures, 120
  marriage, 115, 122
  naturalist, 112
  opposed by Murchison, 119
  professor at Harvard, 121
  professor at Neuchâtel, 115
  Studies of Glaciers, 118
  travel, in Europe, 118
  travel in North America, 120
Agricola, 36
Albert, Prince, 96
Alexander, 6
Algal deposits, 206
American Geological Society, 131
American Journal of Science, 135
Anaximander, 3
Angus, 84
Antiquity of Man (Lyell), 96, 97, 110
Antisell, 217

Arabs, 17
Archaeozoon, 206
Archeozoic, 198
Archimedes (fossil), 170
Aristotle, 4
  Christian authority, 20
  death, 7
  early life, 5
  geologic processes, 7
  marriage, 5
  study under Plato, 5
  tutor of Alexander, 6
  zoological work, 6
Athens, 4
Aufgeschwemmte Gebirge, 44
Auvergne, 31, 64, 66
Averroes, 19
Avicenna, 18

Bacon, Roger, 20
Badlands, 176, 223, **175**
Barrande, Joachim, 107
Basalt, 35
  D'Aubuisson on, 63
  Lyell on, 86
  Von Buch on, 66
  Werner on, 43
Bauer, Georg, 36
Belemnite pen, 21
Black, Joseph, 53
Braun, Alexander, 114

Index

Bruno, 20
Buckland, William, 86, 90, 119
Burnet, Thomas, 22

Cambrian, 104, 107
  conflict, 105
Canadian Shield, 194
  map, 195, 203
Canals, English, 71
  Smith's work on, 72
Carpenter, W. B., 204
Cerauniae, 21
Chamberlin, Thomas Chrowder, 270, 271
  Chinese investigation, 283
  climatic theories, 278
  death, 285
  diastrophism, 285
  education, 271
  glacial epochs, 276
  glaciation, 273, 276
  marine circulation, 279
  Origin of the Earth, 285
  planetesimal hypothesis, 281
  president, Wisconsin, 274
  professor, Beloit, 272
  professor, Chicago, 276
  textbook of geology, 283
  Two Solar Families, 285
  U.S. Geological Survey, 273
  Wisconsin survey, 272
Charpentier, 116
Clarke, John M., 160
Clinton, De Witt, 140
Coal, Logan on, 181
  Owen on, 170
  Volney on, 127
Coastal Plain, 124
Coast and Geodetic Survey, 228
Colonna, 25
Colorado River, explored by Powell, 235
  Powell's report, 238
  Powell's route, 238
Concerning Stones (Theophrastus), 11
Correlation, diagram of, 74
  Smith's methods, 73
Coutchiching, 196
Cuvier, 132
Cycles, Aristotle's theory, 10

Dark Ages, 16

Darwin, 95
  Origin of Species, 95
  traced glaciers, 119
D'Aubuisson, Jean François, 63
Davie, James, 51, 53
Da Vinci, Leonardo, 19
Dawson, J. William, 204, **190**
  Eozoon conflict, 205
De Beaumont, 90
Debris, transportation, 266
Deltas, Aristotle on, 9
Deluge, *see* Flood
Deposition, Guettard on, 30
  Hutton on, 56
  Steno on, 23
  Thales on, 3
Deserts, Powell on, 241
Desmarest, Nicolas, 34
Devonian, 104
De Witt, Benjamin, 132
Driftless Area, 273, 275

Earth, age of, 20, 201, 212, 229, 280
  origin of, 42, 43, 277, 280
Earthquakes, Aristotle on, 7
  Herodotus on, 3
  Strabo on, 12
  Thales on, 3
  Volney on, 127
Earth's crust, theories of, 202
Eaton, Amos, 137, 111
  agent, 138
  county surveys, 144
  death, 149
  earnings, 148
  education, 137
  Erie Canal survey, 145
  Geological Text-Book, 145
  imprisonment, 139
  Index to the Geology of the Northern States, 143, 144
  lawyer, 138
  Lyceum of Natural History, 146
  Manual of Botany, 141
  marriage, 138, 140
  pardon, 140
  popular lectures, 141
  Rensselaer School, 146
  student at Yale, 140
  teaching methods, 147
  Wernerian ideas, 143
  Williams College, 141

Index

Emmons, Ebenezer, 152, 156, **174**
Eozoon canadense, 204, 207, **190**
   conflict, 205
Erie Canal survey, 145
Erosion, Desmarest on, 36
   Guettard on, 30
   Playfair on, 60
   Thales on, 3
Eruptions, Moro on, 25
Erzgebirge, 39
Evans, 217
Evolution, Lyell on, 95
Exhalations, Aristotle's theory, 8

Fall Line, 124, 126
Fallopio, 20
Flood, 20, 22, 42, 45, 90, 144, 209
Flötz Gebirge, 43
Forfarshire, 84
Forster's Chart, 156
Fossil Fishes (Agassiz), 117
Fossils, Anaximander on, 3
   Avicenna on, 18
   Fallopio on, 20
   Guettard on, 30
   Herodotus on, 4
   Leonardo on, 19
   Lister on, 25
   Smith on, 73
   Steno on, 23
   Xanthos on, 4
   Xenophones on, 4
Freiberg, 39

Geognosy, 41, 62
Geological and Geographical Survey of Territories, 226
   map, 227
Geological Text-Book (Eaton), 145, 208
Geology, first used, 41
Geology of Canada (Logan), 187, 192
Gesner, Conrad, 36
Gilbert, Grove Karl, 251, **270**
   chief geologist, 262
   death, 268
   debris, researches, 266
   education, 252
   Great Basin, 263
   Henry Mountains, 256
   keen observer, 251

laccoliths, 257
Lake Bonneville, 259
   marriage, 256
   Ohio survey, 254
   San Francisco earthquake, 266
   U.S. Geological Survey, 256
   Ward's Cosmos Hall, 253
   Wheeler survey, 254
Glacial Period, 117
   divisions, 276
Glacial System (Agassiz), 119
Glaciation, cause, 277, 278
   Chamberlin on, 273
   Logan on, 188
   Von Buch on, 68
Glaciers, Agassiz on, 116
   discovered in United States, 221
   Playfair on, 59
Grand Canyon, 237
Granite, Hutton on, 58
   Von Buch on, 67
Greece, geography, 1
Green River, 235, 238
Grenville, 197
Greywacke, 101
Ground water, Leonardo on, 19
Guettard, Jean Etienne, 27, 37
   childhood, 28
   mineralogic map, 29, 32
   personality, 33
   writings, 33

Hall, James, 150, 174, **174**
   assistants, 158
   basins, theory of, 161
   check to Charles Schuchert, 159
   death, 163
   education, 151
   extent of formations, 161
   financial difficulties, 155, 157
   Forster's Chart, 156
   Geological Survey, Canada, 157
   geologist of Iowa, 158
   glaciers, 153
   manuscript, 160
   metamorphism, 162
   mountain making, 161
   New York survey, 152
   Palaeontology, 154, 160
   personality, 155, 158, 162
   professor of chemistry, 151
   State Paleontologist, N.Y., 154

Index

Hayden, Ferdinand, 223, **255**
   collector of Hall, 224
   death, 228
   education, 223
   exploration in West, 224
   Geological and Geographical Survey of Territories, 226
   professor, 225
   U.S. Geological Survey, 228
Heat, Hutton on, 57
Henry Mountains, Gilbert on, 256
Herodotus, 3
Hot springs, Aristotle on, 8
   Guettard on, 32
Human evolution, Lyell on, 95
Humboldt, 117
Huronian, 185, 192, 199
   glaciation, 199, 201
Hutton, James, 49, **15**
   chemical work, 51, 53
   education, 50
   farmer, 52
   geologic work, 54
   personality, 53
   Theory of the Earth, 55, 56

Ibn-Sina, 18
Ice Age lakes, map of, 265
Illustrations of the Huttonian Theory (Playfair), 59
Index to the Geology of the Northern States (Eaton), 143, 144
Irrigation, Powell on, 242
Islands, Moro on, 25

Jameson, Robert, 62, 101

Keewatin, 193, 197
Keweenawan, 193, 199
King, Clarence, 217, 232, **254**
   California survey, 219
   death, 231
   earth, age of, 229
   education, 218
   Geological Exploration, Fortieth Parallel, 220
   geologic theories, 230
   glaciers of United States, **221**
   metamorphism, 230
   mining engineer, 229
   Systematic Geology, 222, 229

   trip West, 219
   U.S. Geological Survey, 228
Kinnordy, 84
Kirwan, Richard, 56, 143

Laccoliths, 257
Lake Bonneville, 259
   map of, 261, 265
Lamarck, 79, 126
Laplacian theory, 277
   Chamberlain's rejection of, 278
Laurentian, 185, 192, 197
Laurentide Mountains, 188
Lawson, 193, 202
Lehmann, Johann, 42
Leidy, Joseph, 224
Leonardo da Vinci, 19
Lhuyd, Edward, 25
Lister, Martin, 25
Locke, John, 172
Logan, William E., 179, 191, 202, 204, **191**
   business, 180
   death, 190
   education, 179
   exhibitions, 186, 187, 189
   Geological Survey, Canada, 182
   Geology of Canada, 187, 192
   Huronian, 185
   knighted, 187
   Lake Superior region, 184
   Laurentian, 185
   personality, 190
   return to North America, 181
   study of coal, 180, 181, 182
Lowell Institute, 93, 120
Lyell, Charles, 85, 155, **30**
   Antiquity of Man, 96
   basalt, 86
   childhood, 85
   death, 97
   education, 86
   evolution, 95
   glaciers, 87
   in Italy, 89
   lawyer, 87
   lectures, 93
   marriage, 92
   Principles of Geology, 90
   revision, 95, 96
   professor, 92
   Travels in North America, 94

Index

trip with Murchison, 89
visit to North America, 93
Lyell, Charles (elder), 84

Maclure, William, 128, 143, 167, 95
  geologic map, 130
Malesherbes, 31
Man, evolution of, 95
Maps, Guettard's mineralogic, 29, 32
  Maclure's geologic, 130
  Smith's new geologic, 81
  Smith's stratigraphic, 79, 31
Marine circulation, 279
Marl, New Jersey, 130
Meander River, 3
"Metamorphic Series," 185
Metamorphism, Hall on, 162
  Hutton on, 58
Meteorology, Aristotle's treatise, 7
Miletus, 2
Mitchill, Samuel, 132, 133, 138, 140, 110
Moro, Abbé, 25
Moulton, F. R., 281
Mountain making, Hall on, 161
Mountains, Avicenna on, 18
  Colonna on, 25
  Steno on, 24
Murchison, Roderick, 88, 102, 94
  accomplishments, 108
  death, 110
  evolution, 110
  glaciation denied, 109, 119
  knighted, 109
  personality, 109
  Sedgwick and, 102, 105
  Siluria, 105
  Silurian System, 104
  travel, 103
  trip with Lyell, 89
Murray, John, 134

Napoleonic wars, 47, 99
Natural History (Pliny), 11, 13
Neptunists, 61
Newberry, J. S., 217
New Harmony, 167
New York geological survey, 151
  Hall's report, 153
New York System, 153
Niagara Falls, Volney on, 128
Nile, 3

Old Red Sandstone, 103, 104
Ordovician, 107
Origin of Species (Darwin), 95
Origin of the Earth (Chamberlin), 285
Owen, David Dale, 165, **175**
  Arkansas survey, 178
  childhood, 166
  Chippewa Land District, 174
  coal, 170
  death, 178
  education, 166, 168, 169
  geologist, Indiana, 169
  Hall, collecting with, 174
  in Indiana, 168
  Iowa-Illinois-Wisconsin survey, 172
  Kentucky survey, 177
  lectures, 168
  marriage, 169
  museum, 168
  Northwest Territory, 175
  reports, 173, 176
  Smithsonian Institution plan, 174
Owen, Robert, 131, 166

Pacific Railroad surveys, 217
  map of, 218
Palaeontology of New-York (Hall), 154, 160
Persian Empire, 4
Philip II, 6
Piedmont, Volney on, 126
Planetesimal hypothesis, 281
Plato, 5
Playfair, John, 53, 59, 61
Pliny, 12
  methods of work, 12
Powell, John Wesley, 232, **254**
  Army officer, 233
  attacks on, 242
  Colorado River, 235
  death, 250
  deserts, 241
  director, U.S. Geological Survey, 240
  early work, 233
  education, 232
  ethnologic work, 235, 240, 248, 249
  Exploration of the Colorado River, 238
  geologic work, 244
  irrigation, 242

Index

Powell, John Wesley (cont'd)
  lecturer, 233
  marriage, 233
  philosophy, 249
  professor, 234
  Unita Mountains, 244
  U.S. Geological and Geographical Survey, Rocky Mountain Region, 227, 239
  wounded at Shiloh, 233
Pre-Cambrian, map of, 195
  time chart, 200, 201
Primary, 43, 57
  abandoned, 200
Primitive, 43
Principles of Geology (Lyell), 90
  revision of, 95, 96
Prodrome (Steno), 23
Proterozoic, 198
  map of, 203
Pumice, 46
Puys, 31, 35

Raynold's expedition, 225
  map of, 227
Rensselaer School, 146
Richardson, Rev. Benjamin, 75, 76
Rivers, Aristotle on, 9
  Leonardo on, 19
  Strabo on, 12

Salisbury, Rollin D., 272
San Francisco earthquake, 266
Schöpf, Johann, 124
Sedgwick, Adam, 98, 102, 108, 94
  death, 108
  education, 99
  Murchison and, 102
  professor, 100
  quarrel with Murchison, 105
  travel, 100
Sedimentary rocks, Hutton on, 56
Sharks' teeth, 21, 23
Silliman, Benjamin, 133, 110
  American Journal of Science, 135
  death, 136
  lectures on geology, 134
  marriage, 134
  Siluria (Murchison), 105
Silurian, 103, 107
Silurian System (Murchison), 105
Smith, William, 70, **30**

  canal engineer, 72
  card of the English strata, 76, 78
  consulting engineer, 77
  death, 82
  delays in publication, 75, 78
  encouraged by Richardson, 75
  geologic maps, 79, 81, **31**
  geologic work, 72
  lectures, 81,
  publications, 80
  surveyor, 71
Smithsonian Institution, Owen's plan for, 174
Sparta, 4
Steensen, Nils, 22
Steno, Nicolaus, 22
Strabo, 11
Stratified rocks, Moro on, 25
Studies on Glaciers (Agassiz), 118
Sumner, Jack, 234, 239

Tempe, gorge of, 3
Tertiary Period, 91
Thales, 2
Theophrastus, 11
Theory of the Earth (Hutton), 55, 56
Thessaly, 3
Thunder wedges, 21
Time charts, 201, 212, 215
Timiskaming, 198
Tongue stones, 21, 23
Torrey, John, 140, 143
Townsend, Rev. Joseph, 77
Transition, 44, 101
Travels in North America (Lyell), 94
Tuscany, Steno on, 24
Two Solar Families (Chamberlin), 285

Universal formations, 43
U.S. Geological and Geographical Survey, Rocky Mountain Region, 239
  map of, 227
U.S. Geological Survey established, 228
  Powell's directorship, 240
Ussher, James, 20

Van Rensselaer, Stephen, 144, 152
Vesuvius, 12
Victoria, Queen, 96
Volcanoes, Aristotle on, 8
  Desmarest on, 35, 36
  Guettard on, 32

Index

Moro on, 26
Steno on, 24
Strabo on, 12
Volney, Constantin, 124
Volvic, 31, 35
Von Buch, Leopold, 65, 117
  personality, 66, 68, 14
Vulcanism, 134
Vulcanists, 61
Vulkanische Gebirge, 43

Waves, Thales on, 3
Werner, Abraham Gottlob, 39, 14
  dogmatism, 44

influence, 45, 61
personality, 41, 46
teaching, 40
universal formations, 43
Wheeler survey, 223, 226
Whinstone, 58
Whiston, William, 22
Wistar, Casper, 133
Woodward, John, 22

Xanthos, 4
Xenophones, 4

Zirkel, 222